Sources
in the History of Mathematics and Physical Sciences

T0214552

Springer
London
Berlin
Heidelberg
New York
Barcelona
Hong Kong
Milan
Paris
Singapore
Tokyo

Sources in the History of
Mathematics and Physical Sciences

Andersen K.
Brook Taylor's Work on Linear Perspective

Cannon J.T., Dostrovsky S.
The Evolution of Dynamics:Vibration Theory from 1687 to 1742

Chandler B., Magnus W.
The History of Combinatorial Group Theory: A Case Study in the History of Ideas

Dale A.I.
A History of Inverse Probability: From Thomas Bayes to Karl Pearson

Dale A.I.
Pierre-Simon Laplace: Philosophical Essay on Probabilities

Federico P.J.
Descartes on Polyhedra: A Study of the *De Solidorum Elementis*

Goldstein B.R.
The Astronomy of Levi Ben Gerson (1288-1344)

Goldstine H.H.
A History of Numerical Analysis from the 16th through the 19th Century

Goldstine H.H.
A History of the Calculus of Variations from the 17th through the 19th Century

Graßhof G.
The History of Ptolemy's Star Catalogue

Grootendorst A.W.
Jan De Witt's *Elementa Curvarum*

Hermann A., Meyenn K. von, Weisskopf V.F. (Eds)
Wolfgang Pauli: Scientific Correspondence I: 1919-1929

Heyde C.C., et al
I.J. Bienayme: Statistical Theory Anticipated

Continued after Index

Ian Tweddle

Simson on Porisms

An Annotated Translation of Robert Simson's Posthumous Treatise on Porisms and Other Items on this Subject

With 227 Figures

Springer

Ian Tweddle
Department of Mathematics
University of Strathclyde
Livingstone Tower
26 Richmond Street
Glasgow G1 1XH
UK

Sources Editor:
Gerald J. Toomer
2800 South Ocean Boulevard, 21F
Boca Raton, FL 33432
USA

ISBN 978-1-84996-862-1

British Library Cataloguing in Publication Data
Tweddle, Ian
    Simson on porisms : an annotated translation of Robert
    Simson's posthumous treatise on porisms and other items on
    this subject. - (Sources in the history of mathematics and
    physical sciences)
    1. Porisms - History
    I. Title II. Simson, Robert
    516.2'2

Library of Congress Cataloging-in-Publication Data
A catalog record for this book is available from the Library of Congress

© Springer-Verlag London Limited 2010
Printed in Great Britain

Robert Simson (1687–1768)
Portrait by Peter De Nune, 1746
(Courtesy of the Collins Gallery, University of Strathclyde)

Preface

Robert Simson is generally recognised as the first person to have made significant progress in resolving the mysteries of Euclid's lost work on Porisms. He published a paper on this topic in the *Philosophical Transactions* in 1723; but, although he continued to investigate Porisms throughout the rest of his life, he published nothing more on this topic during his lifetime, and it was left to his executor to assemble his various manuscripts into the posthumous *Tractatus de Porismatibus*. This treatise, the paper and various extracts from Simson's notebooks and correspondence form the subject matter of *Simson on Porisms*. My aim has been to produce an accurate translation supplemented by historical and mathematical notes; I hope thereby to add something to the reputation of Simson, who, I believe, has been rather neglected by mathematical historians, possibly because of his obsession with ancient geometry at a time when great strides were being made in contemporary mathematics by mathematicians such as Euler, the Bernoullis and Simson's fellow-countrymen Colin MacLaurin and James Stirling. Simson's claim to fame rests upon his scholarship, which is no less worthy of recognition.

Several people have helped me directly or indirectly with this project. I would like to record my thanks to the following in particular:

(i)   Peter Ottaway, a former undergraduate student, whose need for an honours project in the history of mathematics led to our working on Simson's paper, which in turn inspired me to study the treatise;

(ii) Professor R.A. Rankin of the University of Glasgow for mutual exchange of ideas and information on Simson;

(iii) the Special Collections staff of the Andersonian Library, University of Strathclyde, for help with material and references;

(iv) the staff of the Department of Special Collections at Glasgow University Library for providing me with the bulk of my material;

(v)   Professor G.J. Toomer, the series editor, for his helpful comments and encouragement and, in particular, for saving me from a few gaffs in translation and mathematical interpretation;

(vi) my wife, Grace, and my son, Edward, for their interest and support.

I am grateful to the Department of Special Collections, Glasgow University Library, and to the Library of the Royal Society for permission to quote from manuscripts in their possession. The portrait of Robert Simson appears by kind permission of the Curator of the Collins Gallery, University of Strathclyde. I am likewise indebted to the Librarian of Glasgow University Library concerning the reproduction of four pages from Simson's notebooks.

Finally, it is a pleasure to acknowledge my indebtedness to all the members of the editorial and production staff at Springer with whom I have dealt.

IT
Glasgow
April 2000

Contents

Introduction

    Background 1

    Some Remarks on the Work 5

    Mathematical Requirements, Terminology, Notation 6

Simson's Title 9

Part I. Introductory Material

Summary 11

Preface 13

Definitions 17

Propositions 1–6 18

Pappus's Account of the Porisms 33

Notes on Part I 38

Part II. Pappus's Two General Propositions and Euclid's First Porism

Summary 45

Propositions 7–25 47

Notes on Part II 91

Part III. Lemmas and Restorations

Summary 97

Propositions 26–79 99

Notes on Part III 188

Part IV. Various Porisms: Fermat, Simson and Stewart

Summary 199

Propositions 80–93 201

Notes on Part IV 242

Appendices

A1. A Translation of Simson's 1723 Paper along with some
    Comments 251

A2. "That this goes to a given point" 263

A3. Correspondence between Pappus's Lemmas and
    Simson's Propositions 265

A4. Corrections to Simson's Text 266

References 269

Index 273

Introduction

Background

Robert Simson (1687–1768) was Professor of Mathematics at the University of Glasgow from 1711 until his retirement in 1761. He had originally studied for the ministry at Glasgow but mathematics, which he first came to as a relaxation from theological arguments, gradually became his dominant interest. During 1710–11 he spent about a year in London pursuing his mathematical studies and becoming acquainted with several eminent mathematicians, in particular Edmund Halley, who was largely responsible for Simson's dedicating himself to the study of Greek geometry.

Early on Simson became intrigued by Euclid's *Porisms*. What was known about these was outlined by Pappus of Alexandria in Book 7 of his *Mathematical Collection* and became known to Simson through the edition of Commandino.[1] According to Pappus the three books of Porisms have thirty-eight Lemmas and there are one hundred and seventy-one Theorems. However, although he gave statements of theorems from all three books, most of these are in very contracted form and have become mutilated, with the result that only the first Porism recorded from the first Book and his statements of two general propositions, one incorporating ten loci from the first Book, the other a proposed generalisation of the former, have come down to us intact. It is not even clear from Pappus what kind of result should be regarded as a Porism. He gives one definition which he attributes to geometers who came after Euclid and who, he asserts, did not understand the true nature of Porisms. He then gives a definition which he attributes to the ancients; but it is so general as to be almost all-embracing. In 1706 Halley published his Greek and Latin translations, made from Arabic manuscripts, of Apollonius's *The Cutting off of a Ratio* [13], a work also discussed by Pappus in Book 7, and included with it a translation of Pappus's description of the Porisms; Halley concluded that Pappus's account would be of little value to anyone trying to understand the nature of the Porisms.

After much frustrated effort punctuated by attempts to close his mind to the Porisms, Simson finally made a break-through in the early 1720s. The

---

[1] Federicus Commandinus Urbinas [5]. Halley presented Simson with a copy of this work which contained Halley's own notes.

story is recounted in both William Trail's *Account of the Life and Writings of Robert Simson, M.D.* [40] and Lord Brougham's *Lives of Philosophers of the Time of George III* [2] of how Simson, walking one day (probably in April 1722) on the banks of the River Clyde with some friends, became detached from his companions, fell into a reverie concerning the Porisms, was struck by some new ideas and, having drawn a diagram with his chalk on a nearby tree, first penetrated the mysteries of the Porisms. On 22 March 1723 Simson communicated to Dr Jurin, Secretary of the Royal Society, some of the fruits of his labours [47]:

> The Inclosed contains the Restitution and Demonstration of the two Propositions in Pappus relating to the Porisms which however easy they may appear now they are found out yet have oftner than once baffled my keenest Endeavours to discover their meaning so far as to make me despair of ever obtaining it ....

The "two Propositions" are Pappus's general propositions mentioned above; Simson added the first two Porisms recorded from Book I by Pappus. According to his letter he had at that time much more material, including his conjectures for the ten individual loci which had been incorporated in the first of Pappus's general propositions. Simson's paper was subsequently published in the *Philosophical Transactions*.[2] He was of course delighted at this outcome and wrote to Dr Jurin on 10 January 1724 [47]:

> The Honour you have done me in presenting the Paper I sent up in the Philo: Trans: is what I am very sensible of; and you may be sure the approbation any thing in it has had from such good Judges as Dr. Halley and Mr. Machin, yourself and the other Learned Gentlemen you were pleas'd to show it to, cannot but give me a great deal of Pleasure; and nothing could excite me more to Endeavour to Restore the other Porisms, of which there are the least Data: But I find it a very difficult affair especially to one who is so slow, as by much experience I find myself: ....

Simson appears to have persevered in this endeavour, but he published nothing more on Porisms during his lifetime. In his will [48] he bequeathed to his friend and former colleague James Clow, Professor of Philosophy at Glasgow University, all his manuscripts "earnestly desiring & requesting him that he will put the same into proper order and cause print & publish the same in whole or in part as he shall judge fitt ... " With financial assistance from the Earl Stanhope, an amateur mathematician and friend of Simson's,

---

[2] 'Pappi Alexandrini Propositiones duae generales, quibus plura ex Euclidis Porismatis complexus est, Restitutae a Viro Doctissimo Rob. Simson, Math. Prof. Glasc. Vid. Pappi praefationem ad Lib. 7. Coll. Math. Apollonii de Sectione rationis libris duobus a Clariss. Hallejo praemissam pag. VIII & XXXIV' [26]. I have provided a translation of this paper in Appendix 1. The reference in the title is to Halley's work [13] mentioned above; the first page reference is to the Greek translation, the second to the Latin.

Clow produced in 1776 the *Opera Quaedam Reliqua* [32], an edition of some of Simson's manuscripts, the second item of which is listed on the title page as "Porismatum liber, quo doctrinam hanc veterum geometrarum ab oblivione vindicare, et ad captum hodiernorum adumbrare constitutum est" and subsequently as "De Porismatibus Tractatus"; this *Treatise on Porisms*, which occupies almost two hundred and eighty pages of the *Opera Quaedam Reliqua*, is the testament to Simson's work on Porisms[3] and forms the principal subject matter of the present volume. Simson's manuscript appears to be lost, but his *Adversaria*, a set of sixteen notebooks preserved in Glasgow University Library,[4] contain many entries relating to Porisms, some of which are clearly drafts of material for his paper or *Treatise*.

Simson's paper in the *Philosophical Transactions* was well received. In particular it was responsible for launching his former student Colin MacLaurin on his very successful investigations in what is effectively projective geometry; MacLaurin wrote: "I was led into these new theorems by Mr Robert Sympson's giving me at that time a Hint of the ingenious Paper, which has been since published in the Philosophical Transactions" [21]. Why did Simson not capitalise on this initial success, especially when he appears to have had no lack of material? According to Trail [40, p. 23]

> There are many indications of his intention of publishing the Porisms; but from various causes he postponed the execution of it, till, in the progress of life, he acquired so very strong an impression of the decline of his faculties, that he reluctantly gave up the design.

Certainly in the early 1740s Simson regarded the publication of his work on Porisms as a matter of some urgency and importance (see the note on Simson's Preface, p. 38). Even in the early 1760s he had not entirely abandoned the idea: in Volume Q of the *Adversaria* there is a draft of a title page for his restoration of Apollonius's *Determinate Section*[5] which lists the Porisms as an addendum; neighbouring dated entries suggest that this draft was written in 1762. There are various references to Simson's being lazy and devoid of ambition,[6] but I believe that these were only apparent traits caused by his diffidence and that the real reason for Simson's delaying was at least in part

---

[3] Trail asserts: "I may take the liberty of adding that though Dr. Simson's fame, as an accurate, elegant, and ingenious Geometrician, be established from his other works; yet the restoration of the Porisms of Euclid will be regarded by posterity as the most important production of those powers of investigation and genius, with which he was so eminently endowed." [40, p. 52]

[4] See [43]. It is curious that some of the entries have been written in decreasing page order.

[5] In fact this was not published in Simson's lifetime but appeared as the first item, "Apollonii Pergaei de Sectione Determinata Libri II restituti, duobus insuper libris aucti"[33], in the *Opera Quaedam Reliqua*.

[6] On 17 April 1738 Simson's colleague Francis Hutcheson, Professor of Moral Philosophy, wrote to the Rev. Thomas Drennan in Belfast: "Robt. Simson, if he were not indolent beyond imagination, could in a fortnights application finish another book which would surprise the connoiseurs." Then on 5 August 1743 he wrote

that he had an entrenched fear of failing to do justice to the great figures of the past. This is well illustrated in the case of his restoration of the *Plane Loci* of Apollonius [27]: according to Lord Brougham, Simson's text was completed in 1738, but it was 1746 before he agreed to publication; he then had second thoughts, stopped copies being issued and bought back those which had already been sold; a further three years elapsed before he finally released the work (see also footnote (6)). Nevertheless, Simson appears to have exchanged material on Porisms with his students to their mutual advantage, notably with Matthew Stewart, who became Professor of Mathematics at Edinburgh University in 1747 following MacLaurin's death.[7]

An English translation of Simson's *Treatise* was begun almost immediately by John Lawson:[8] some forty pages ending just after the start of the proof of Simson's Proposition 17 were published at Canterbury in 1777 under the title *Robert Simson: A Treatise concerning Porisms, translated from the Latin by John Lawson B.D.* [35], but nothing more appeared. In November 1847 Robert Potts of Trinity College, Cambridge, launched a proposal to have a translation of "Dr Simson's Restoration of the Porisms of Euclid" published by subscription at Cambridge University Press; by the end of 1849 there was a substantial list of subscribers, among whom were many distinguished mathematicians, but the proposal seems to have gone no further.[9] It is interesting to note that Simson's ideas were taken up enthusiastically on continental Europe: a German edition based on Simson's text by August Richter appeared in 1837 [25] and the great French geometer Michel Chasles, while not always agreeing with Simson's interpretations or restorations, followed Simson's approach in his work on Porisms [3].

With the publication in 1986 of Alexander Jones's excellent account of Pappus's Book 7 this important text has become generally accessible [17]. Jones makes several references to Simson's work, in particular to his work on Porisms. It seems therefore that an English translation of Simson's *Treatise*

---

again to Drennan: "... we expect immediately from Robt. Simson a piece of amazing Geometry, reinventing 2 Books of Appollonius, & he has a third almost ready. He is the best Geometer in the world, reinventing old Books, of which Pappus preserves only a general account of the subjects." The third book is presumably the Porisms. (See [49].)

[7] Further details of the interplay between Simson and Stewart will be found in the note on Simson's Preface, p. 38.

[8] John Lawson (1723–1779) was a lecturer and tutor in mathematics at Sidney Sussex College, Cambridge, and became Rector of Swanscombe, Kent, in 1759. Among his several mathematical publications the most notable is perhaps *A Dissertation on the Geometrical Analysis of the Antients, with a Collection of Theorems and Problems without Solutions* [18], which generated a great deal of interest. The dissertation was reproduced as the first article in the first volume of Thomas Leybourn's *The Mathematical Repository* [19] and many of the unproved propositions were stated and subsequently proved in several articles by a variety of contributors.

[9] A copy of this proposal with the list of subscribers is held in the University of London Library – see [23].

is long overdue and would be of value to modern scholars of Greek geometry. This I have tried to provide in *Simson on Porisms*.

Some Remarks on the Work

I have resisted the temptation to recast the material in modern form and language and have tried to retain the spirit of Simson's text. However I have indulged in a certain amount of redrafting in cases where a direct English translation seemed excessively cumbersome or obscure.

The diagrams are closely based on those contained in Simson's text. Where Simson has provided a single composite diagram to cover several cases of the main result, a converse, corollaries, extensions, etc., I have usually extracted a series of figures for the individual parts.

The original text is annotated with marginal notes for reference to other results and footnotes for extended remarks; these are referenced in alphabetical sequence or by *, †, ‡ (the latter may be later additions). For ease in setting, the marginal notes have been included in the body of the text, enclosed in square brackets and usually located at the end of the items to which they refer. The footnotes are still set as such and I have retained Simson's identifiers even where the alphabetical sequence has been broken as a result of the incorporation of the marginal notes within the text; thus, for example, the first footnote of Proposition 89 is indicated by *f*. There are three places (Propositions 47, 91) where marginal marks (*, †) serve to locate items for subsequent reference; these have been retained.

Following standard practice, Simson refers to results in Euclid's *Elements* only by item and Book number (e.g. 19.5 identifies the 19th item in Book 5 of the *Elements*). I have changed such references by putting the Book number first in Roman numerals followed by the item number in ordinary numerals; thus Simson's 19.5 becomes V 19. The abbreviation *Dat.* identifies Euclid's *Data*. It is important to note that Simson's references are to his own editions of the *Elements* and the *Data*,[10] since his numbering varies occasionally from that found in other editions due to his addition or reordering of material. Fortunately the results cited are usually self-evident and elementary, so that it has not seemed necessary to expand on Simson's references except in a few cases.

Simson's text is not broken down into sections in any formal way, although he does indicate in his Preface that the results have been put together in specific categories. I have retained Simson's ordering of the material but I have presented the translation in four parts, each of which is introduced by a brief statement of its contents and is concluded with a series of notes on

---

[10] The second edition of Simson's *The Elements of Euclid* [31] includes *The Book of Euclid's Data*.

individual results or sections. I have identified the presence of a note by putting the page number of the note in the left-hand margin at the start of the item to which it applies. The notes contain both mathematical and historical comment. A note on a Proposition may contain some discussion relating to a specific point in its demonstration; the location of the part of the text and the relevant discussion in the note are identified by (n) (or (n1), (n2), ... if the note contains several such items) at appropriate places in the text and the note. However, Simson's discussions are generally quite clear and consequently some items have no accompanying notes.

Simson made many additions to material which he quoted from Pappus; such items are usually identified by italics or by the use of square brackets or quotation marks. In a few places I have made additions to Simson's text – these are contained within angled brackets.

Mathematical Requirements, Terminology, Notation

While Simson establishes many elegant propositions, his demonstrations are generally elementary, although often lengthy, and require only the basic rules for manipulating ratios, the elementary properties of ratios associated with similar triangles or intersections of lines with systems of parallel lines and some basic properties of circles and cyclic quadrilaterals; Menelaus's Theorem from plane geometry also finds application.

Following standard practice (cf. Jones's *Pappus* [17]) I have not attempted to translate the technical terms referring to the manipulation of ratios. Those used by Simson are as follows.

If $a : b = c : d$ then

$b : a = d : c$ (*invertendo*),

$a : c = b : d$ (*permutando*, also called *alternando*),

$(a + b) : b = (c + d) : d$ (*componendo*),

$(a - b) : b = (c - d) : d$ (*dividendo*, also called *separando*),

$(b - a) : b = (d - c) : d$ (*dividendo inverse*),

$a : (a - b) = c : (c - d)$ (*convertendo*).

If $a : b = c : d$ and $b : e = d : f$ then $a : e = c : f$ (*ex aequali*).

If $a : b = c : d$ and $b : g = h : c$ then $a : g = h : d$ (*ex aequali in proportione perturbata*).

Simson also makes use of the facts that if $a : b = c : d$ then $(a + c) : (b + d) = a : b$ and, where meaningful, $(a - c) : (b - d) = a : b$.

The term *compound ratio* is used for a product of ratios.

Simson's demonstrations usually follow the standard process of Greek geometry and consist of the following:

(i) an *analysis*, in which the proposition is assumed to be true and a conclusion which is known to be true is reached by a series of deductions;

(ii) a *composition* or *synthesis*, in which steps of the analysis are reversed to establish the proposition.

In some cases these two aspects are formalised, in others Simson passes from one to the other without comment.

By the *rectangle* AB, CD Simson means the product of the lengths AB and CD. Signs are never involved, so that AB and CD are always non-negative quantities and AB and BA are used interchangeably.

Finally, we should note Simson's practice of separating off subsidiary parts of an argument within a continuous text by means of parenthesis; for example, in Propositions 5 we have: "Therefore the rectangle AH, DE is equal (to the rectangle F, DE, that is, as was shown, to the rectangle AB, BD, twice the rectangle DB, BC, and the square of BC; that is to the rectangle AB, BD, and the rectangle DB, BH and the square of BC; that is) to the rectangle AH, DB and the square of BC." Here the conclusion is that $AH.DE = AH.DB + BC^2$; the reasoning is contained within the parenthesis.

Roberti Simson, M.D.
De
Porismatibus
Tractatus;
Quo
Doctrinam Porismatum Satis Explicatam,
Et in Posterum ab Oblivione Tutam
Fore sperat Auctor.

A Treatise on Porisms
by
Robert Simson, M.D.

Whereby the author hopes that the doctrine of Porisms
will be explained sufficiently and will be safe from oblivion
in the future.

Part I. Introductory Material

We begin with Simson's Preface, in which he describes the background to his work and outlines its contents. This is followed by a set of definitions, which includes a statement of what Simson understood by a Porism. Next Propositions 1–6 provide simple illustrative examples. Finally we have Pappus's description of the Porisms, which Simson has taken from Halley.

Preface

Nothing about the Porisms apart from what Pappus saved is found in the writings of the ancient Geometers. And Fermat asserts in *Var. Oper. Math.* p. 116 that the modern writers perhaps did not even know them by name, or only suspected what they were. However I find that before Fermat, Albert Girard had thought about their explanation and restoration: in his *Trigonometry*, written in French and published at The Hague in the year 1629, after the enumerated forms of straight-line figures which have four, five or six sides, he adds these things, "When there are only two lines which pass through a point, the whole thing is like the lost Porisms of Euclid once were; these I hope to bring to light soon, having restored them some years ago." And again, in the *Mathematical Works of Simon Stevin*, which Girard published in amended and extended form at Leiden in 1634, after he had said on page 459 that Euclid quite rarely made use of a compound ratio, he adds, "But it may be assumed that he wrote more about it in his three books of Porisms which are lost; God willing, I hope to bring them to light, having rediscovered them." From these things it is clear that Girard, a man certainly uncommonly skilled in mathematics, thought that he had restored the Porisms; however what he had written about them was never published, and, unless by chance it is lying hidden in some library in Holland, it has to be considered as entirely lost. Moreover, from the first quotation it seems that he regarded as Porisms certain propositions about quadrilateral and other figures, and their relationships, and consequently that he did not understand their nature.

After Girard, Ismael Boulliau tried to explain the Porisms in his third *Exercitatio Geometrica*, which along with the first two was published at Paris in the year 1667; but although he had acquired certain things about them from Fermat, for he quotes Fermat's words, which were afterwards printed at Toulouse in the year 1679 in the restored doctrine of Porisms among his various mathematical works, Boulliau could not elucidate them at all. Moreover, in this book of Fermat's, which was published after his death, there are just a few things, but, it seems, everything that he had put down in writing about the doctrine of Porisms. From these it is clear that this very acute man was the only one since the time in which Pappus lived to whom something about their nature had become known. But since he acknowledges that he had penetrated into the secrets of this matter with almost no other help than that provided by the words of the faulty definition of Porism which Pappus rightly blames the more recent Geometers for having given, namely that a Porism is that which is lacking in hypothesis from a local theorem, it is certain that he had not sufficiently understood what the Porisms are. For in fact Fermat asserts quite erroneously that this definition specifically reveals the nature of a Porism, for there are innumerable Porisms which in no way depend upon a local theorem and have nothing in common with loci. Fermat does indeed promise in the same work that he would restore all three books of the Porisms at some time or other and would reveal other wonderful and unknown ones (these are his own words); however these words are not a little rash, for there are many of Euclid's Porisms of which not the slightest trace remains; yet Fermat did not even explain the first Porism of the first book, which is the only one preserved in its entirety by Pappus.

There is no point in repeating the things that Carlo Renaldini has on this topic in his book *De Resolutione et Compositione Mathematica*, for they contribute in no way to the understanding of Porisms. And these are the only authors, as far as the modern writers are concerned, who have mentioned Porisms, at least those who in some way have tried to explain them. And although the most distinguished David Gregory in his preface to *The Works of Euclid* deemed that "it will not be difficult to restore the Porisms to some extent where the Greek text of Pappus sheds light", his colleague, the most learned Halley, a man somewhat more experienced in the Geometry of the ancients than Gregory, following on his edition of the Greek text amended as far as possible, adds forthrightly to Pappus's description of the Porisms the following comment, "The description of the Porisms given up to this point was neither understood by me nor will it be of use to the reader, and it could not be otherwise, both on account of the lack of the figure which is mentioned, as a result of which quite a lot of lines, with which we are concerned here, are jumbled together without alphabetical notation or any other distinguishing character, and also because certain things have been omitted and transposed, or otherwise corrupted, in the exposition of the general proposition; hence what Pappus intended is in no way given to me to conjecture. Add to these things the extremely abbreviated mode of expression, which is not suitable for use in a difficult thing such as this is."

However, after I had read in Pappus that Euclid's Porisms were a very ingenious collection of many things which are concerned with the analysis of more difficult and general problems, I was possessed by a great desire to know something about them; consequently I tried repeatedly and in many different ways to understand and restore both Pappus's general Proposition, which was maimed and imperfect, and also the first Porism of Book I, which, as has been said, is the only one in all of the three books to remain intact up to this time; but in vain, for I made no progress. And since thoughts about this matter had consumed much of my time, and had eventually turned out to be quite troublesome, I firmly resolved never to investigate these things in future, especially as the excellent Geometer Halley had abandoned all hope of understanding them. Hence as often as such thoughts came into my mind, so I turned them away. However it happened afterwards that they came upon me when I was not expecting them and was forgetful of my intention and kept me occupied until eventually a certain light shone forth which engendered in me the hope of finding at least Pappus's general Proposition, which indeed after much investigation I finally restored. In fact this Proposition along with the first Porism of Book I was published a little later in the *Philosophical Transactions* No. 177 for the year 1723.

But since at that time I had not sufficiently penetrated the nature of the Porisms, I now want to hand it over in a more detailed fashion, so that this class of propositions and the way in which they are investigated, which have lain hidden from Geometers from the time in which Pappus lived up to the present, may be restored to Geometry again, and may bring to it a not unprofitable addition. And since Pappus's description of the Porisms cannot easily be understood without an example of them, so it seemed appropriate for the explanation of those which Pappus recorded to begin with certain simple Porisms and then to add to Pappus's description certain of Euclid's Porisms, namely those which I was able to distinguish as his, either from Pappus's general proposition or from his description of the Porisms, or else finally with the help of his lemmas for the Porisms. These are followed by four of Fermat's propositions transformed into the form of Porisms, for the remaining one of his five is about the Parabola and we have explained it in Proposition 19 of Book 5 of *Sect. Conic.* (2nd edition). Certain others have been added. The eminent Geometer Matthew Stewart, Professor of Mathematics in the University of Edinburgh, proposed the most important of these to me and gave the construction of some of them. This subject has already been excellently cultivated by him and should, as I hope, be much cultivated by him later.

One may hope therefore that by these things the doctrine of Porisms is sufficiently explained and will be safe from oblivion in the future.

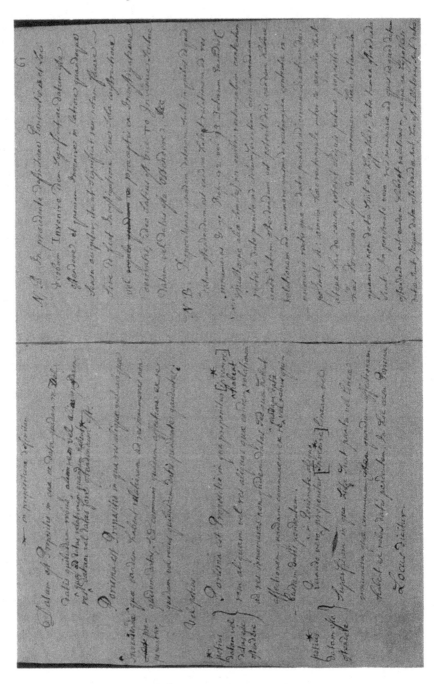

Simson's definition of Porism
Adversaria, Vol. E, pp. 60–61
(Courtesy of Glasgow University Library)

On Porisms:
Definitions

1. A *Theorem* is a Proposition in which something is proposed for demonstrating.

2. A *Problem* is a Proposition in which something is proposed for constructing or for finding.

3. A *Datum*, or a Proposition about given things, is a Theorem in which it is proposed to show that something is given (according to the definitions of Euclid's Data) which has a certain stated relationship to those things that are given by hypothesis.

A Datum can also be enunciated in the form of a Problem, if indeed those things which have to be shown to be given, are proposed for finding. Because if this is done, the demonstration of the Datum, set forth as a Theorem, will be the Analysis of the Problem; moreover the Composition corresponding to the Analysis will be the Construction and Demonstration of the Problem.

Since Pappus's definition of a Porism is exceedingly general, in place of it let there be the following, viz.

4. A *Porism* is a Proposition in which it is proposed to show that something or several things are given when it is required that a certain common property described in the Proposition is satisfied by it or by them as well as by any one of innumerable things which are not in fact given but which have the same relation to those things that are given.

A Porism can also be enunciated in the form of a Problem, if indeed those things which have to be shown to be given, are proposed for finding.

5. A *Locus* is a Proposition in which it is proposed to show given or to find the curve or surface, each point of which has a certain common property described in the Proposition, or to show given or to find the surface in which any curve given by a stated rule has such a property.

Whence it is clear, as Pappus asserts, that Loci are a form of Porism; and the common property which is proposed for proving about these points or curves, is that they all are located on one particular curve or surface, which has indeed to be found.

Furthermore, other Propositions in which it is proposed to show or to find something, apart from those which are Data or Porisms, are simply called Theorems or Problems.

(p.42)   # Proposition 1

"If the straight line AB is given in position, and the circle CD is given in position and magnitude, then the point with the following property will be given: if any straight line is drawn through it to meet the straight line AB and the circle, then the rectangle contained by the segments of that straight line between the point to be found and the points of intersection will be given."

Lemma

Let there be a triangle ABC having a right angle at A; if BC is bisected at D and AD is joined, then BD will be equal to DA. This is the converse of the first part of Proposition 31 of Euclid's Book 3.

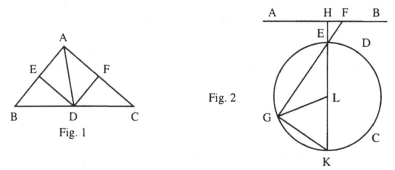

Fig. 2   Fig. 1

For (Fig. 1) let DE be drawn parallel to AC and DF parallel to AB. The angle DBE is therefore equal to the angle CDF, as also is the angle BDE to the angle DCF, and the sides BD, DC, which are adjacent to the equal angles, are equal; therefore the straight line BE is equal to the straight line DF, that is, to EA. And in triangles BED, DEA the side DE is common, and the angles at E are right angles, for the angle at A is a right angle. The base BD is therefore equal to the base DA. This also follows from the Corollary to Proposition 5 of Euclid's Book 4.

Now (Fig. 2) let E be the point which has to be found in the Porism, and, any straight line EF having been drawn through E, let it meet the straight line AB in F, and the circle in G; therefore, by hypothesis, the rectangle FE, EG is given. Let the perpendicular EH be drawn to AB, and let it meet the circumference in K, and let KG be joined; therefore, again by hypothesis, the rectangle HE, EK is given, which consequently is equal to the rectangle FE, EG. Thus as FE is to EH, so EK is to EG, and they are about equal angles, so that the angle KGE is equal to the right angle EHF [VI 6]. Let EK be bisected in L, and let LG be joined; therefore LG is equal to LK [Lemma]. Likewise all the straight lines drawn from the point L to the circumference will be shown equal to the same LK. The point L is therefore the centre of the circle CD; and LE is equal to LK, so that the point E is on the circumference of the given circle. But the point L is given [Dat. Defn. 6], therefore the straight line LH is given in position, and the points E, H, K will be given;

consequently the rectangle HE, EK and the rectangle FE, EG, which is equal to it, are given.

It will be composed as follows.

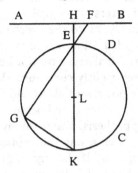

From the centre L of the given circle let LH be drawn perpendicular to AB, and let it meet the circumference in points E, K; either of these two points, say E, will satisfy the Porism. For let any straight line FEG be drawn through E, and let GK be joined; therefore since the angle EGK in the semicircle is equal to the right angle EHF, the triangles EGK, EHF are equiangular; therefore as GE is to EK, so EH is to EF, and consequently the rectangle GE, EF is equal to the given rectangle KE, EH.

In this Porism it has to be shown that the point E is given, and the innumerable things which have the same relation to the point E, and the given straight line AB, and the given circle CD, are the segments of the straight lines between the point E and the straight line AB, and the circle CD. Now the common property which it is proposed to show satisfied by the same segments is that the rectangle contained by them is given; this rectangle also had to be found.

The converse of this Porism is the following Locus, viz. Suppose that from a given point E two straight lines EF, EG are drawn in the same straight line, so that they contain a given area; moreover let the end F of one lie on a straight line AB which is given in position; then the end G of the other will lie on the circumference of a circle which is given in position. This Locus is Proposition 8 of Book 1 of Apollonius's *Plane Loci*, which were published at Glasgow in the year 1749. Now the Theorem can clearly be derived from this Locus, on account of which the ancients called it a Local Theorem, that is, if from the centre L of a circle the perpendicular LH is drawn to a straight line AB and meets the circumference in E, K, and if through the point E any straight line FEG is drawn which meets the straight line AB in F and the circumference again in G, then the rectangle FE, EG will be equal to the rectangle HE, EK. However if the hypothesis of this Theorem is reduced, or diminished by one condition, that is, if it is not stated in it that it is the point E through which the straight lines are to be drawn, but it is set forth in the following way, viz. Suppose that a straight line AB has been given in position and that a circle CD has been given in position and magnitude; then the point will be given which has the property that if any straight line is drawn through it which meets AB and the circumference, then the rectangle contained by the segments of the straight line drawn between the point to be found and the points of intersection will be given; it is clear that the proposition has now been changed into the preceding Porism. Therefore this example clearly explains the definition of the younger Geometers by which they say that a Porism is what is derived from a Local Theorem by reducing the hypothesis. Pappus indeed rightly condemns this definition because it is based on something which is not essential, as will be shown in what follows.

From this example it is also clear that Loci are also derived from a Local Theorem by reducing the hypothesis. For if the hypothesis of the preceding Local Theorem is reduced in such a way that nothing is said in it about the circle, but only the following, viz. If through a given point E any straight line FEG is drawn whose other end F lies on a straight line which is given in position, and if the rectangle FE, EG is equal to a given area, and it is proposed to find the curve on which the other end G lies, it is clear that the Theorem is now converted into a Locus. Whence Pappus asserts that Loci themselves are a form of this type of Porism, namely that to which the definition of the younger Geometers applies.

(p.42)
Proposition 2

"Suppose that a point is given and a circle is given in position and magnitude; if a straight line is drawn from any point on the circumference to the given point, another point will be given with the property that if a straight line is drawn to it from the same point on the circumference, the line first drawn will have a given ratio to this line."

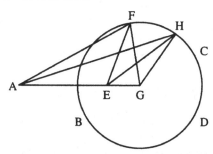

Let the given point be A, and the given circle be BCD; suppose that the Porism is true, and let E be that point which it is required to find. Therefore if from any point F on the circumference FA, FE are drawn, then, by hypothesis, the ratio which AF has to FE will be given. Let AE be joined, and let FG be drawn to AE produced to make the angle EFG equal to the angle EAF; therefore since triangles GAF, GFE are equiangular, AG will be to GF and GF will be to GE as AF is to FE; therefore AG, GF, GE are proportional, consequently the straight line AG is to the straight line GE as the square of AG is to the square of GF [VI 20, Corollary 2]; whence the square of AF is to the square of FE as AG is to GE. Again let any other point H be taken on the circumference, and let HA, HE, HG be joined. Therefore since, by hypothesis, the ratio AH to HE is the same as the ratio AF to FE, the square of AH will be to the square of HE as the straight line AG is to the straight line GE; therefore, by the Lemma to Proposition 2 of Book 2 of Apollonius's *Plane Loci* in the edition which was printed in Glasgow, which is Proposition 119 of Pappus's Book 7, the rectangle AG, GE is equal to the square of GH. And since AG, GF, GE are proportional, the same rectangle AG, GE is equal to the square of GF; consequently the straight line GH is equal to the straight line GF. And likewise any straight line drawn from the point G to the circumference may be shown equal to the same GF. Therefore the point G is the centre of the given circle, and consequently it will be given; and the point A is given, consequently the straight line AG is given, and it will have a given ratio to GF, which is given in magnitude [*Dat.* 1]. Moreover the ratios AF to FE and GF to GE have been shown to be the same as this ratio, so that those ratios will be given. And since GF is given, as is its ratio to GE, then GE will be given in magnitude [*Dat.* 2]; but it is also given in position, and the point G is given, therefore the point E is given.   Q.E.D.

It will be composed as follows.

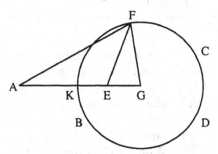

Let A be a given point, let G be the centre of a given circle, and AG having been joined let it meet the circumference in K; now let a third proportional to AG, GK be found, namely GE. The point E will be that point which was to be found, and the ratio AG to GK will be that ratio which was to be shown given. For if from the points A, E straight lines AF, EF are drawn to meet on the circumference in any manner, the ratio AF to FE will be the same as the ratio AG to GK. Let FG be joined, and since AG, GK, or GF, and GE are proportional, triangles GAF, GFE will be equiangular [VI 6]; thus AG is to GF, that is to GK, as AF is to FE.

In this Porism the point E was to be shown given or to be found, and the innumerable straight lines which can be drawn from the points A, E to meet on the circumference have the same relationship to the points A, E and the circumference. Moreover this common property, which had to be shown to be satisfied by these straight lines, is the following, viz. these straight lines have among themselves a given ratio, which was also to be found.

The converse of this Porism is the Locus which is contained in Proposition 2 of Book 2 of Apollonius's *Plane Loci*, from which the following Local Theorem follows, viz. Suppose that the straight line AG is drawn from a point A to the centre G of a circle BCD, and to AG and the semidiameter GK there be taken a third proportional GE which is located on the straight line GA towards A; then if straight lines AF, EF are drawn in any manner from the points A, E to meet on the circumference, they will have the same ratio as the ratio AG to GK. Now if that part is removed from the hypothesis of this Theorem which says how the point E is to be found, and what the ratio of the straight lines is, then, as is clear, the Theorem will be changed into the preceding Porism.

Proposition 3

"Suppose that two points A, B are given on a straight line and that a ratio is given; on the same straight line let two other points C, D be taken which make the ratio AC to BD the same as the given ratio; another point will be given which makes the ratio of the segments between itself and the points C, D given."

A   C   E   D B   E   A   B C   D

Fig. 1   Fig. 2

Suppose that the Porism is true, and let E be that point which makes the ratio EC to ED given. Therefore since A, B are given points, if E is also given, the straight lines EA, EB will be given in magnitude and consequently their ratio will also be given. Therefore since in Fig. 1. the ratios of the whole of AE to the whole of EB, the part AC to the part BD, and the remaining part CE to the remaining part ED are given, and in Fig. 2 the ratios of the whole of EC to the whole of ED, the part AC to the part BD, and the remaining part AE to the remaining part EB are given, the ratios of all segments to all other segments will be given [*Dat.* 12], if of course the given ratios are not the same. Therefore if they are not the same, the ratio of AE to each of CE, ED will be given; and since AE is assumed given, CE, ED will be given, and since the point E is assumed given, the points C, D would be given. However they are not given; for C and D are any points which make the ratio AC to BD the same as the given ratio. Therefore the ratios CE to ED, AC to BD, and AE to EB are the same. And since the ratio AC to BD is given, the ratio AE to EB, which is the same as it, will be given. And AB is given, so that AE is also given, and the point E will be given.

It will be composed as follows.

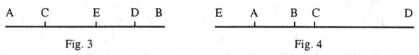

Fig. 3   Fig. 4

If the points C, D are on opposite sides of the points A, B (Fig. 3), let the point E be taken between the points A, B, so that it makes the ratio AE to EB the same as the given ratio; if however the points C, D are on the same side of A, B (Fig. 4), let the point E be taken in AB produced in the direction away from C, D, so that it makes the ratio AE to EB the same as the given ratio. In either case CE will be to ED as AE is to EB. For since, by construction, AE is to EB as the given ratio which AC has to BD, then CE will be to ED as AE is to EB [V 19 or V 12].

The same thing in a different form.

A   C F   E GD B   E   A   B F C   G D

Let E be the point to be found which makes the ratio EC to ED given. Now let two other points F, G be taken which make the ratio AF to BG the same as the ratio AC to BD; therefore, by hypothesis, the ratio EF to EG will be the same as the ratio EC to ED. Therefore since AF is to BG as AC is to BD, then CF will be to DG as AC is to BD [V 19 or V 12]; and since EF is to EG as EC is to ED, then CF will be to DG as EC is to ED [V 19 or V 12], so that EC is to ED as AC is to BD; and consequently AE is to EB as AC is to BD [V 19 or V 12]. Moreover the ratio AC to BD is given, therefore the ratio AE to EB is given; and AB is given, so that the point E is given. And it was shown that EC is to ED as AC is to BD; therefore the ratio EC to ED is given. The composition is the same as in the preceding version.

Proposition 4

"If the line AB is given in position and magnitude, and any point C is taken on it produced, then the rectangle AC, CB along with a certain given area, will be equal to the square of the segment between the point C and a certain given point."

F   A   D   B   E   C

Suppose that the Porism is true, and let the area which is to be shown given or is to be found, be called S; and let D be the point which has to be found. Therefore, by hypothesis, the rectangle AC, CB along with the area S is equal to the square of DC. Again let any other point E be taken on AB produced, and, by hypothesis, the rectangle AE, EB along with the area S will be equal to the square of DE. Therefore the excess of the rectangle AC, CB over the rectangle AE, EB will be equal to the excess of the square of DC over the square of DE. If furthermore AF is made equal to BE, the excess of the rectangle AC, CB over the rectangle AE, EB is equal [a] to the rectangle FC, CE. Now the excess of the square of DC over the square of DE is twice the rectangle DE, EC along with the square of EC [II 4], which are both consequently equal to the rectangle FC, CE. Let the square of EC be taken away from both sides, and twice the rectangle DE, EC will be equal to the rectangle FE, EC. Therefore twice DE is equal to FE; consequently FD is equal to DE, and their segments FA, BE are equal; therefore the remaining part AD is equal to the remaining part BD. Moreover AB is given, consequently AD is given, and the point D will be given. Now the rectangle AC, CB along with the area S is equal to the square of DC, that is to the rectangle AC, CB along with the square of DB [II 6]; therefore the area S is equal to the square of DB. And consequently the area S is given.

It will be composed as follows.

A   D   B   C

Let AB be bisected in D; then D will be the point, and the square of DB will be the area which are required to be found. For let any point C be taken on AB produced; the rectangle AC, CB along with the square of DB will be equal to the square of DC (n), which is in accordance with II 6 or can be established from the preceding analysis as follows: since the rectangle AC, CB is equal to the rectangle AB, BC along with the square of BC [II 3], let the square of DB be added to both, and the rectangle AC, CB along with the square of DB will be equal to the rectangle AB, BC, that is to twice the rectangle DB, BC, along with the squares of BC, BD, that is to the square of DC [II 4].

(p. 42)

Proposition 5

"Three points A, B, C having been given on a straight line, let any point D be taken between A, B; on the straight line AC a point will be given which will make the rectangle contained by the segment between it and the point D, and a certain given straight line equal to the rectangle AD, DB along with the square of the segment between D and the third point C."

A   G   D   B   E   C   H

F

---

[a] This is Proposition 24 of Pappus's Book 7 . Let AB be bisected in D, and since the rectangle AC, CB along with the square of DB is equal to the square of DC, that is, to the rectangle FC, CE along with the square of DE, that is, along with the rectangle AE, EB and the square of DB. Let the square of DB be taken away from both sides; therefore the remaining rectangle AC, CB will be equal to the rectangle FC, CE, and the rectangle AE, EB.

Suppose that the Porism is true, and let E be the point to be found and let F be the straight line to be found. Therefore since, by hypothesis, the rectangle F, DE is equal to the rectangle AD, DB along with the square of DC, that is to the rectangle AD, DB, the squares of DB, BC and twice the rectangle DB, BC [II 4], then the rectangle F, DE will be equal to the rectangle AB, BD, twice the rectangle DB, BC, and the square of BC [II 3]. Again let any other point G be taken between A and B; and it is shown similarly that the rectangle F, GE is equal to the rectangle AB, BG, twice the rectangle GB, BC, and the square of BC. Therefore the excess of the rectangles F, GE, F, DE, that is the rectangle F, GD, will be equal to the excess of the rectangles AB, BG, AB, BD along with twice the excess of the rectangles GB, BC, DB, BC, that is to the rectangle AB, GD along with twice the rectangle BC, GD. Therefore the straight line F is equal to the straight line AB along with twice BC, that is to the straight lines AC, CB together; therefore if CH is made equal to CB, the straight line F will be equal to the straight line AH, and consequently the straight line F is given. Therefore the rectangle AH, DE is equal (to the rectangle F, DE, that is, as was shown, to the rectangle AB, BD, twice the rectangle DB, BC, and the square of BC; that is to the rectangle AB, BD, and the rectangle DB, BH and the square of BC; that is) to the rectangle AH, DB and the square of BC. Let the rectangle AH, BD be taken away from both sides, and the remaining rectangle AH, BE will be equal to the square of BC. And BC, AH are given, so that BE is given, and the point E will be given. And the line F is given, as has been shown. Q.E.D.

It will be composed as follows.

A   D   B E   C   H

F

Let BC be produced to H so that BC, CH are equal, and let BE, the third proportional to AH, BC lying on AB produced, be found. Then AH will be the straight line called F which has to be found and E will be the point which has to be found. For let any point D be taken between A and B. Now the rectangle AH, BE is equal, by construction, to the square of BC; moreover the rectangle AH, DB is equal (to the rectangle AB, BD along with the rectangle HB, BD, that is, because BH is twice BC,) to the rectangle AB, BD along with twice the rectangle DB, BC; thus the sum, namely the rectangle AH, DE will be equal (to the sum of the rectangle AB, BD, twice the rectangle DB, BC, and the square of BC, that is to the rectangle AD, DB, the squares of DB, BC, and twice the rectangle DB, BC, that is) to the rectangle AD, DB and the square of DC.

Some people may consider that the analysis can be carried out more succinctly by taking the point D not arbitrarily but from among the given points; and in this case the analysis may be carried out as follows.

Suppose that the Porism is true, and let E be the point and F the line which have to be found. Therefore in the case where the point D coincides with A, since there is no rectangle contained by the segments between D, or A, and the points A, B, it follows, by the hypothesis of the Porism, that the rectangle F, DE, or F, AE, will be equal to the square of AC. Again let the point, which was G in the preceding analysis, coincide with B, and since also in this case there is no rectangle AG, GB, it follows similarly, by hypothesis, that the rectangle F, BE will be equal to the square of BC. Therefore the excess of the rectangles F, AE, F, BE, which is in fact the rectangle F, AB, is equal to the excess of the squares of AC, BC, that is to the square of AB and twice the rectangle AB, BC. Therefore the straight line F is equal to the straight line AB along with twice the straight line BC, that is to the straight line AH, where CH has been made equal to BC. And since the rectangle F, BE, that is AH, BE, is equal to the square of BC, and the straight lines AH, BC are given, BE will be given and consequently the point E will also be given.

But although this way of finding the straight line AH and the point E is shorter than the preceding method, one must be completely on guard against this type of analysis; for a proof of the general Porism cannot be derived from it, as must be done if an arbitrary point D is taken, but only of the particular case concerning the points A and B which were used in the analysis. Therefore at least once the point, the line, the angle, or whatever it is about which something is asserted generally in the Porism, has to be taken arbitrarily, but within the limits prescribed in the Porism, if there are any, in order that the analysis may be carried out properly. However the things which have to be investigated will often be found more succinctly and more easily, if first of all one of the stated things is taken arbitrarily and then starting afresh it is chosen in a certain special way, if this can be done. Thus in the previous Porism, when the point D was taken arbitrarily, it was shown that the rectangle F, DE is equal to the rectangle AB, BD, twice the rectangle DB, BC, and the square of BC; if now in place of the point G which in the initial analysis was taken arbitrarily, the point B is taken, the rectangle F, BE will be equal, by hypothesis, to the square of BC; and when these quantities have been taken away from the former, the remaining parts will be equal, that is, the rectangle F, DB will be equal to the rectangle AB, BD and twice the rectangle DB, BC. Therefore the straight line F is equal to the straight line AB along with twice BC and consequently the straight line F is given; and the point E will be shown to be given as in the first analysis, whose composition also serves for this approach.

K   A   D   B L E M C   N   H

F

Figure for Corollaries 1–4

Corollary 1. Suppose that in place of the point D which was taken between the points A, B, a point such as K is taken in AB produced on the side of A; the rectangle F, KE, or AH, KE, along with the rectangle AK, KB will be equal to the square of KC.

For the rectangle AH, BE is equal, by construction, to the square of BC; and the rectangle AH, KB is equal (to the rectangle KB, BA along with the rectangle KB, BH, that is, because BH is twice BC,) to the rectangle KB, BA along with twice the rectangle KB, BC; when these equalities have been added, the rectangle AH, KE will be equal to the rectangle KB, BA, twice the rectangle KB, BC and the square of BC; let the rectangle AK, KB be added on both sides, and the rectangle AH, KE along with the rectangle AK, KB will be equal to the square of KB, twice the rectangle KB, BC, and the square of BC, that is to the square of KC.

Corollary 2. And if the point is taken between B and E, such as L, then in this case also the rectangle AH, LE along with the rectangle AL, LB will be equal to the square of LC.

For since the rectangle AH, BE is equal to the square of BC, when the rectangle AL, LB has been added to both sides, the rectangle AH, BE along with the rectangle AL, LB will be equal (to the square of BC along with the rectangle AL, LB, that is to the squares of BC, BL along with the rectangle AB, BL, that is,) to twice the rectangle CB, BL, the square of LC, and the rectangle AB, BL [II 7]; and the rectangle AH, BL is equal to the rectangle AB, BL, (along with the rectangle BH, BL, that is) along with twice the rectangle CB, BL. And when these equal quantities have been taken away from the former, the rectangle AH, LE along with the rectangle AL, LB will be equal to the square of LC.

Corollary 3. When the point, such as M, is between E and C, the rectangle AM, MB will be equal to the rectangle AH, ME along with the square of MC.

For the rectangle AM, MB along with the square of BC is equal (to the rectangle AB, BM along with the squares of BM, BC, that is to the rectangle AB, BM, twice the rectangle CB, BM, and the square of MC [II 7], that is, because BH is twice BC, to the rectangle AB, BM along with the rectangle HB, BM, and the square of MC, that is) to the rectangle AH, BM along with the square of MC; of these terms the square of BC is equal to the rectangle AH, BE; therefore the remaining rectangle AM, MB is equal to the remaining part, namely the rectangle AH, ME along with the square of MC.

Corollary 4. When the point is taken in AC produced, such as N, the rectangle AN, NB will be equal to the rectangle AH, NE along with the square of NC.

This is shown in the same way as Corollary 3.

Finally if the point A is taken in place of D, the rectangle AH, AE will be equal to the square of AC. If the point B is taken, the rectangle AH, BE will be equal to the square of BC. And if it is taken at E, the rectangle AE, EB will be equal to the square of EC. And if the point C is taken, the rectangle AH, CE will be equal to the rectangle AC, CB.

Moreover these can all be illustrated by numbers, by taking, for example, 24 for AB, 12 for BC; for then F, or AH, will be 48, BE will be 3. And if 10 is taken for AD, then F, DE will be 816, AD, DB will be 140, and the square of DC will be 676, and so on in the remaining cases.

Proposition 6

"Suppose that three points A, B, C are given on a straight line of which C lies between A and B; a point D will be given on AB produced on the side of A, such that if a point E is taken arbitrarily on AB produced on the same side, the rectangle BC, CE will be equal to the rectangle DA, AE along with the rectangle contained by the straight line EB, and a certain given straight line."

Suppose that the Porism is true, and let D be the point and F the straight line which have to be found. Therefore, by hypothesis, if the point E is taken arbitrarily on BA produced, the rectangle BC, CE will be equal to the rectangle DA, AE along with the rectangle F, EB. Now let the point A be taken in place of E, and again, by hypothesis, the rectangle BC, CA will be equal to the rectangle F, AB. And when these quantities have been taken away from the preceding equal quantities, the remaining rectangle BC, AE will be equal to the remaining part, namely the rectangle DA, AE along with the rectangle F, AE. Therefore the straight line BC is equal to the straight lines DA, F together; consequently the rectangle AB, BC is equal to the rectangle DA, AB along with the rectangle F, AB. Now it was shown that the rectangle BC, CA is equal to the rectangle F, AB, and when these have been taken away from the immediately preceding equal quantities, the remaining part, namely the square of CB, will be equal to the remaining rectangle DA, AB. But BC is given and so the rectangle DA, AB is given; and AB is given, consequently the straight line AD and the point D are given. But BC is equal to the straight lines DA, F together; therefore the line F is given.

It will be composed as follows.

E   D   A   C·   B   F
——————————————————————   ——

Let AD, the third proportional to AB, BC taken from A on BA produced, be found. Further let a straight line F be taken equal to the excess of the straight lines CB, AD, that is, let CB be equal to AD, F together; therefore the square of BC is equal to the rectangle DA, AB; let the rectangle AC, CB be added to both, and the whole rectangle AB, BC will be equal to the sum, namely the rectangle DA, AB along with the rectangle AC, CB. And since the straight line CB is equal to the straight lines AD, F together, the rectangle AB, BC will be equal to the rectangle DA, AB along with the rectangle F, AB. Therefore the rectangle DA, AB along with the rectangle AC, CB is equal to the rectangle DA, AB along with the rectangle F, AB; consequently the rectangle AC, CB is equal to the rectangle F, AB. Again since the straight line BC is equal to the straight lines AD, F together, the rectangle BC, AE will be equal to the rectangle AD, AE along with the rectangle F, AE; when these have been added to the immediately preceding equal quantities, the whole rectangle BC, CE will be equal to the sum, namely the rectangle DA, AE along with the rectangle F, EB.   Q.E.D.

E   D   A G C   H   B K   F
——————————————————————   ——

Figure for Corollaries 1–3

Corollary 1. If in place of the point E the point G is taken between A and C, the rest remaining as before, the rectangle DA, AG along with the rectangle BC, CG will be equal to the rectangle F, GB.

For since the rectangle DA, AB is equal to the square of CB, let the rectangle GC, CB be added to both, and the rectangle DA, AB along with the rectangle GC, CB will be equal to the rectangle GB, BC. But the straight line CB is equal to the straight lines DA, F together, consequently the rectangle GB, BC, that is the rectangle DA, AB along with the rectangle GC, CB is equal to the rectangle DA, GB along with the rectangle F, GB. Let the rectangle DA, GB be taken away from both, and the rectangle DA, AG along with the rectangle GC, CB will be equal to the rectangle F, GB.

Corollary 2. If the point H is taken arbitrarily between C and B, the rectangle DA, AH will be equal to the rectangle BC, CH along with the rectangle F, HB.

For since the straight line CB has been shown to be equal to the straight lines DA, F together, the rectangle HB, BC will be equal to the rectangle DA, HB along with the rectangle F, HB. Let the rectangle DA, AH be added to both, and the rectangle DA, AH along with the rectangle HB, BC will be equal (to the rectangle DA, AB along with the rectangle F, HB, that is,) to the square of CB along with the rectangle F, HB. Let the rectangle HB, BC be taken away from both, and the rectangle DA, AH will be equal to the rectangle BC, CH along with the rectangle F, HB.

Corollary 3. If the point K is taken arbitrarily on AB produced, the rectangle DA, AK along with the rectangle F, KB will be equal to the rectangle BC, CK.

The rectangle DA, AB is equal to the square of CB, and the rectangle DA, KB along with the rectangle F, KB is equal to the rectangle CB, BK, because the straight line DA along with the straight line F is equal to the straight line CB; thus the first sum, the rectangle DA, AK along with the rectangle F, BK, will be equal to the other sum, the rectangle KC, CB.

Porisms can also be proposed about numbers; for example, the preceding Porism can be stated for numbers as follows.

If two numbers a, b have been given whose larger is a, two other numbers x, y will be given such that, if any fifth number c is taken, the product of the excess of the first a over the second b, and the sum of the second b and the fifth c, is equal to the product of the third x, and the fifth c, along with the product of the sum of the first a and the fifth c, and the fourth y.

Therefore by hypothesis there will be

$$\overline{a-b} \times \overline{b+c} = cx + ay + cy$$
$$\text{i.e. } ab + ac - bb - bc = cx + ay + cy.$$

Again let any number d be taken in place of c, and it is shown similarly that

$$ab + ad - bb - bd = dx + ay + dy;$$

and if c is greater than d, when the latter equation has been subtracted from the former, there will be

$$a \times \overline{c-d} - b \times \overline{c-d} = \overline{c-d} \times x + \overline{c-d} \times y;$$

therefore $a - b = x + y$. Whence $a - b - y = x$, and $ac - bc - cy = cx$. And there was

$$ab + ac - bb - bc = cx + ay + cy;$$

therefore

$$ab + ac - bb - bc = ac - bc - cy + ay + cy.$$

Whence

$$ab - bb = ay, \quad \text{and} \quad \frac{ab - bb}{a} = y.$$

And there was $a - b - y = x$, consequently

$$a - b - \frac{ab - bb}{a} \quad \text{i.e.} \quad \frac{aa - 2ab + bb}{a} = x = \frac{\overline{a-b}^2}{a}.$$

Or more briefly thus.

It is shown as before that

$$ab + ac - bb - bc = cx + ay + cy,$$

therefore in the case where $c = 0$, there will be $ab - bb = ay$, and when these have been taken away from the preceding quantities, there will be $ac - bc = cx + cy$. Whence there will be $a - b = x + y$, and $a - b - x = y$, and $aa - ab - ax = ay$. But there was $ab - bb = (ay =)aa - ab - ax$; therefore

$$\frac{aa - 2ab + bb}{a} \quad \text{i.e.} \quad \frac{\overline{a - b}^2}{a} = x.$$

For example, if $a = 9$, $b = 3$, then $x = 36/9 = 4$ and $y = 2$, and if any number is taken for c, for example unity, then $\overline{a - b} \times \overline{b + c} = 6 \times 4 = 24$ and $cx = 4$ and $\overline{a + c} \times y = 10 \times 2 = 20$ and $24 = 4 + 20$.

But if $c = 10$, the rest remaining as before, then $\overline{a - b} \times \overline{b + c} = 6 \times 13 = 78$, $cx = 40$, $\overline{a + c} \times y = 38$ and $78 = 40 + 38$.

.42)   It is clear from these examples that there are many Porisms which are derived from a Local Theorem by reducing the hypothesis, but there are others that in no way depend on Loci. It was therefore with justification that Pappus reproved the younger Geometers because they defined a Porism *ex accidente*, namely according to a certain thing which applies to certain but not to all Porisms. Fermat gave the following definition of a Porism on p. 118 of his *Var. Oper. Math.*, viz. "When we investigate a Locus, we search for the straight line or curve at the time unknown to us, until we have marked out the Locus of the curve which is to be found, but when from the supposition that the Locus is given and known we hunt for another Locus, that new Locus is called by Euclid a Porism, for which reason Pappus added that Loci are and are called very correctly one form of Porism. We have supplemented our definition with a single example: in the 5th figure of the Porism, the straight line RC having been given, if any curve RAC is sought with the property that the perpendicular AD from any point A on it makes the square of AD equal to the rectangle RDC, we will find the curve RAC to be the circumference of a circle; but if now, having been given that Locus, we search for another, for example the Problem of the 5th Porism, this new Locus along with the infinitely many others which the acuteness of the experienced analyst will instantly produce, extracting them from what is now known, is called a Porism.

Moreover since we have now said that Porisms are themselves Loci etc."

But what Porisms are cannot be understood from this definition and in fact what Fermat contributes either in his definition or to supplement it, applies to no Porisms other than those which are Loci. For Loci, whether they have been derived from another Locus or from no other Locus, are recognised by Pappus and Fermat himself as a form of Porism, namely they are of that type of Porism to which the definition of the younger Geometers applies; whence it is clear that there are others or at least one other form of this type which are not Loci, apart from other types, which have nothing in common with Loci, about which, it would appear, Fermat knew nothing. Indeed he confuses Porisms with Loci, since he calls his 5th Locus a Porism, and afterwards asserts that Porisms are in general Loci. Moreover the example of the Locus on the circumference from which he wishes his Porism to be deduced, is irrelevant, and has no special connection with it (the Porisms). And indeed it would have been better if Fermat had said that the 5th Porism is the converse of the Locus on page 7 of his book or that it is derived from the Local Theorem arising from that Locus by reducing the hypothesis; and so by this example he would have illustrated the definition of the junior Geometers.

Pappus's Account of the Porisms

Now that these things have been set forth, what Pappus set down in the preface to Book 7 of the *Mathematical Collections* can be more easily understood; this is as follows.

From the Seventh Book

of the Mathematical Collections

of Pappus of Alexandria

"Euclid's Three Books on Porisms"

After the Tangencies Euclid's Porisms are given in three books, forming a very ingenious collection of many things which look at the analysis of more difficult and general problems, of which nature certainly provides a vast supply. Indeed nothing has been added to what Euclid first wrote, except that certain unskilled people who preceded us attached inferior descriptions to a few of them [that is to say, Porisms]. Although each one has a fixed number of demonstrations, [a] as we have shown, [b] Euclid set down one and that the most obvious one in each case. Moreover they have a fine and natural discussion, which is necessary and extremely general, and quite pleasing to those who have the ability to examine and discover individual things. But as a type all of these are neither theorems nor problems, but in a certain manner they are of a nature intermediate between these, so that their propositions can be expressed either as theorems or as problems. Whence it has come about that among many Geometers some consider these to be theorems in type, while others regard them as problems, looking only at the form of the proposition. Now it is clear from the definitions that the ancients had a better understanding of the differences among these three. For they said that a Theorem is that by which something is proposed to be demonstrated; a Problem, that by which something is proposed to be constructed; and a Porism, that by which something is proposed to be discovered. But this definition of Porism has been changed by later writers, who could not investigate all these things, but, applying these Elements, only showed what it is that is sought, but did not investigate it. And however much they were contradicted by the definition and by the very things which have been handed down, nevertheless they made the definition, by something inessential, in this way. A Porism is that which is lacking in hypothesis from a Local Theorem [that is to say, a Porism is a Local Theorem lacking or reduced in its hypothesis]. Moreover Geometrical Loci, of which there is a great supply in the Books on Analysis, are members of this class of Porisms, and items collected separately from the Porisms are presented under their own titles, because this form is more

---

[a] Perhaps he means the proofs of various cases of the same Porism; or of propositions which are its converses.

[b] In certain of the lemmas to the Porisms, unless I am mistaken.

diffuse and abundant than others. For among Loci there are certain that are plane, certain that are solid, certain that are curved and apart from these there are mean Loci [or Loci arising from mean proportionals]. It is also a feature of Porisms that they have concise propositions on account of the difficulty of many things that are usually taken for granted; whence it has come about that not a few Geometers examine the matter only in part, while they understand not at all those things that are more essential among what has been shown. Moreover there is no way in which many of those things can be contained in one proposition, because Euclid himself put only a few things in each form, but in order to give an example of the great abundance, he put a few at the beginning of the first book which are all of the same form as that very abundant form [which is found in the first book], the Loci, so as to enumerate ten. Thus noticing that these [that is to say, the propositions of this form] can be brought together in one proposition, we shall describe it thus.

"Suppose that in a quadrilateral whose opposite angles are either facing or are located in the same direction, [a] [when the sides have been produced] three points [that is to say, points of intersection] are given in one of them; or suppose that in a quadrilateral with two parallel sides [two points of intersection are given in one of the parallel sides;] suppose further that all but one

---

[a] The first of these Pappus calls ὕπτιος, the other παρύπτιος; and it seems that ὕπτιος, that is to say, the figure, is to be understood as a quadrilateral in which two sides AD, CD are inclined in a direction opposite to that in which the other two sides AB, BC are inclined, that is they are inclined back from the first pair. And the figure for παρύπτιος is a quadrilateral in which two sides AD, CD are inclined beside the other two sides AB, CB or in the same direction.

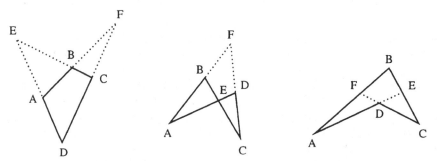

Thus any figure of four sides, no two of which are parallel, is contained in these two figures.

In certain of Pappus's lemmas for the Porisms there is mention of τὸ παράλληλον, namely the figure which is a quadrilateral with two of its sides parallel, as ABCD in the figures below.

of the remaining points lie on a line[b] given in position; then the remaining point will also lie on a straight line given in position." Now this is asserted about only four straight lines, no more than two of which pass through the same point. But in the case of an arbitrary number of straight lines it is not noted to what extent a proposition of the following type may be true, viz.

"Suppose that an arbitrary number of straight lines meet each other, there being no more than two through the same point; suppose further that all the points on one of them are given and each individual point on another lies on a straight line given in position." Or more generally, as follows. "Suppose that an arbitrary number of straight lines meet each other with no more than two through the same point, and let all the points [that is to say, points of intersection] on one of them be given; the number of the remaining points will be a triangular number, whose side expresses the number of points which lie on a straight line given in position; then if no three of these intersections are at the angles of a triangular area [no four at the angles of a quadrilateral, no five at the angles of a pentagon, etc., i.e., generally, if none of these intersections form an orbit] each one of the remaining intersections will lie on a straight line given in position."

Now it is hardly likely that Euclid did not know that this is true, but he was concerned with only the principles: for throughout all the Porisms, apart from first principles, he seems to have sown only the seeds of many great things. Moreover these things are to be distinguished not in terms of differences of hypotheses, but rather according to differences of what results and what is sought. Certainly all the hypotheses differ among themselves, since they are very particular: but each one of the results and questions, since it is one and the same, touches upon many different hypotheses.[a]

And so such things are presented for investigation in the propositions of the first book: (in the beginning of the seventh there is a diagram concerning this[*])

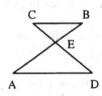
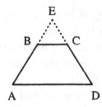

[b] I.e., one lies on one straight line given in position, another lies on another straight line given in position, and so on.

[a] E.g. There are many Porisms which have different hypotheses, but which all conclude that some point lies on a straight line given in position; or that some straight line passes through a given point. etc.

[*] It no longer exists.

"If from two given points two straight lines are inflected to a straight line which is given in position, and if one of them cuts off from a straight line which is given in position a segment adjacent to a given point on it, then the other will also take off from another straight line a segment having a given ratio." i.e., *which will have to the other segment the same ratio as the ratio which is given by the hypothesis.*
Then in what follows:

"That that point lies on a straight line given in position."

"That its ratio . . . to the straight line . . . is given."

"That its ratio . . . to the part cut off is given."

"That this straight line . . . is given in position."

"That this goes to a given point."

"That its ratio is given . . . to the intercept between the point . . . and a given point."

"That the ratio is given of the straight line . . . to some . . . drawn. . . from a point."

"That the ratio of the rectangle ** to the rectangle contained by a given . . . and . . . ."

"That one side of this rectangle is given, while the other has a *given* ratio to the straight line cut off."

"That this rectangle either alone or together with a certain given area is ** that has a given ratio to the part cut off."

"That the straight line . . . along with another to which . . . is in a given ratio, has a given ratio to the intercept between the point . . . and a given point."

"That what is contained by a certain given . . . and the straight line . . . is equal to what is contained by the other given . . . and the intercept between the point . . . and a given . . . ."

"That the ratio is given of the straight line . . . and also of its . . . to the intercept between the point . . . and a given one."

"That the straight line . . . takes off from those given in position segments containing a given rectangle."

In the second book the hypotheses are certainly different. But the things to be investigated are on the whole the same as in the first book; in addition there are the following:

"That that rectangle . . . has a *given* ratio to the part cut off, either by itself or after a certain given rectangle has been added."

"That the ratio is given of the rectangle contained . . . and . . . to the part cut off."

"That the ratio is given of the rectangle contained by both . . . and . . . taken together, and by both of them . . . and . . . also taken together, to the part cut off."

"That what is contained by it ... and by both of them ... and ... which to the straight line ... has a given ratio; and also what is contained ... and by that which has a given ratio to it are in a given ratio to the part cut off."

"That the ratio is given of both ... taken together to the intercept between the point ... and a given point."

"That the rectangle is given which is contained by them ... and ...."

In the third book most hypotheses are about semicircles but there are a few about circles and segments. In fact a very great part of what is to be investigated is connected with what has gone before. In addition the following things present themselves:

"That the ratio is given of the rectangle ... in ... to the rectangle ... in ...."

"That the ratio is given of the square ... to the part cut off."

"That the rectangle contained by them ... and ... is *equal* to the rectangle contained by a given and the intercept between the point ... and a given point."

"That the square of ... is *equal* to what is contained by a given ... and the intercept between the perpendicular and a given point."

"That the straight lines ... along with that to which ... it has a given ratio, taken together, have a given ratio to the part cut off."

"That some point is given with the property that if straight lines are drawn from it any points ... they will contain a triangle given in form."

"That some point is given with the property that if straight lines are drawn from it to any points ... they will cut off equal arcs *from a circle*."

"That a straight line ... will be parallel to a straight line given in position, *or will pass through a given point*, or will contain a given angle with a certain other straight line directed towards a given point."

Furthermore the three books of Porisms have thirty-eight Lemmas and one hundred and seventy-one Theorems.

This is what Pappus recorded.

Notes on Part I

Note on Simson's Preface (p. 13). The texts to which Simson refers are in order of citation:

(i)   Pierre de Fermat, *Varia opera mathematica* .... *Accesserunt selectae quaedam ejusdem epistolae, vel ad ipsum a plerisque doctissimis viris ... de rebus ad mathematicas disciplinas aut physicam pertinentes scriptae* (Tolosae, 1679);

(ii) Albrecht Girard, *Tables des sinus, tangentes & secantes* .... *Avec un traicté succinct de la trigonométrie, etc.* (second edition) (La Haye, 1629);

(iii) Albrecht Girard, *Les oeuvres mathématiques de Simon Stevin* (Leiden, 1634);

(iv) Ismael Boulliau, *Ismaelis Bullialdi exercitationes geometricae tres. I. Circa demonstrationes per inscriptas & circumscriptas figuras. II. Circa conicarum sectionum quasdam propositiones. III. De porismatibus. Astronomiae Philolaicae fundamenta clarius explicata, & asserta aduersus clariss. viri Sethi Wardi ... impugnationem.* (Parisiis, 1657);

(v)   Carlo Renaldini (Count), *Carlo Renaldini ... de resolutione et compositione mathematica ... libri* II (Patavii, 1668);

(vi) David Gregory, *Euclidis quae supersunt omnia. Ex recensione Davidis Gregorii* (Oxford, 1703);

(vii) Robert Simson, *Sectionum conicarum libri* V (second edition) (Edinburgh, 1750).

The four Porisms of Fermat which Simson includes are contained in Propositions 80, 81, 83 and 85 of the *Treatise.* The one which is omitted and which is generalised in the second edition of Simson's *Conic Sections* asserts in Simson's version that "if from two points on a parabola two straight lines are inflected to any third point on the parabola and meet any diameter, and if the diameter meets the straight line through the points from which the straight lines are inflected, then the segments of the diameter between its vertex and the inflected lines will have the same ratio as the segments of the straight line through the points between the points and the point in which it meets the diameter." Simson also discussed this result in several entries in his *Adversaria* and extended it to the same or opposite branches of a hyperbola.[11] Fermat presented his Porisms without giving any demonstrations in his "Porismatum Euclidæorum Renovata Doctrina, & sub forma Isagoges recentioribus Geometris exhibita", which is contained in *Varia opera mathematica*, pp. 116–119.

Matthew Stewart (1717–1785) has already been mentioned in the Introduction. Correspondence between Simson and Stewart [20, 46] from the 1740s

---

[11]See Vols. D, H of [43] and Proposition 20 of [28].

shows that Simson sought Stewart's collaboration in his work on Porisms and that he regarded its publication as an urgent matter at that time. On 27 December 1742 Simson wrote to Stewart, who was then studying under MacLaurin in Edinburgh:

> I renew the desire I made to yow when last here that yow would not communicate these things till I see yow at least. In the mean time I wish yow would as your leasure can permit find out as many as may be of them, and be sure to write down both analysis and composition, because when I publish an account of the Porisms I shall be glad to have your store to encrease mine which shall every one of them be particularly acknowledged in the book.

Stewart was delighted and flattered by Simson's proposal and indicated in his reply of 30 December 1742 that he expected to have "sixteen, if not twenty" entirely new results. He had in fact been intending to publish some of his results independently but not as Porisms:

> I was not resolv'd to publish any of my Propositions in the form of Porisms, tho that was the form I thought several of them ought to appear in, but as you was pleas'd to let me know that you had some thoughts of publishing something on this subject soon I was resolv'd as much as possible to avoid any thing that would seem to have the least tendency to this.

Simson's next letter to Stewart is dated 3 January 1743. In it he makes various suggestions for the publication of Stewart's results – an Appendix to the *Plane Loci*,[12] incorporation in the proposed work on Porisms, or some in one and some in the other – and reiterates his concern for confidentiality:

> As to the keeping private what relates to the porisms I am obliged to your prudence which I hope yow will continue for reasons will satisfie yow fully at meeting. In the mean time yow will think it reasonable, how much soever I encline to communicate any small things I have that may be of any use to the publick, that I should be the first publisher myself; and this has made me keep the porisms to myself, except the few hints Mr Ja: Moor, John Williamson[13] and yourself got from me.

---

[12] See the Introduction, p. 4.

[13] James Moor and John Williamson were both students of Simson. Moor (1712–79) became Librarian at Glasgow University in 1742 and Professor of Greek in 1746. Like Stewart he assisted Simson in the preparation of the *Plane Loci*. In particular Moor checked the Greek text as given by Halley against manuscripts in Paris (see Simson's Preface to the *Plane Loci*). It is perhaps of interest to note that in 1748 Moor purchased in Paris a 16th-century copy of a Greek manuscript of the *Mathematical Collection*, Books 3–6 and 8 (see [41, p. 204]). He was also the author of the following inscription which was placed at one time below Simson's portrait: "Geometriam, sub tyranno barbaro saeva servitute diu squalentem, in libertatem et decus antiquum vindicavit unus." (Roughly: He alone restored to freedom and its former glory Geometry, which had long been squalid from cruel enslavement

Simson also indicated a proposed time scale for the work in a reference to "the account I am to give of the Porisms which will be some small time after the Loci plani are published, I mean within a year or thereabout."

Stewart spent several weeks with Simson about May 1743 helping him prepare the *Plane Loci* for the press. Thereafter he returned to Rothesay, Isle of Bute, where his father was a minister. Probably under the influence of his father he then prepared for and underwent trials for the ministry, becoming licensed to preach in May 1744. The direct link between Simson and Stewart had been broken and this presumably contributed to the further delay in the publication of the *Plane Loci* and Simson's failure to publish his work on Porisms. Stewart, however, went on to to publish his *General Theorems of considerable use in the higher parts of Mathematics* [36], the success of which was largely instrumental in his being elected to the Chair of Mathematics at Edinburgh in September 1747. Simson and Stewart obviously remained on good terms and, although the proposed major contribution to Simson's work by Stewart did not materialise, as Simson notes, the last four propositions (90–93) originated with Stewart. Simson made a favourable reference to Stewart's *General Theorems* in the *Plane Loci* [27, p. 223].

Stewart also published "Pappi Alexandrini collectionum mathematicarum libri quarti propositio quarta generalior facta, cui propositiones aliquot eodem spectantes adjiciuntur" [37], from which Proposition 93 comes, and *Propositiones Geometricae more veterum demonstratae* [38].

Note on Simson's definition of Porism (p. 17). The essence of Simson's idea of Porism seems to be contained in the following scheme:

(i)   there are certain given (fixed) geometrical objects;
(ii)   there are certain proposed geometrical objects;
(iii) there is an infinite collection of geometrical objects defined in terms of those in (i) and (ii);
(iv)   it has to be shown that the objects in (ii) are given when the objects in (iii) are required to satisfy a further specified condition.

The Latin text of Simson's definition of Porism is as follows:

> Porisma est Propositio in qua proponitur demonstrare rem aliquam, vel plures datas esse, cui, vel quibus, ut et cuilibet ex rebus innumeris, non quidem datis, sed quae ad ea quae data sunt eandem habent rationem, convenire ostendendum est affectionem quandam communem in Propositione descriptam.

Volume E of Simson's *Adversaria* contains a draft of the introductory material for the *Treatise* and includes on p. 61 the following variant of the above definition:

---

under a barbarous despot.) Williamson, whom Lord Brougham described as "a favourite pupil and a man of great promise", became Chaplain to the British Factory at Lisbon, where he died young.

Porisma est Propositio in qua proponitur invenire rem aliquam vel res aliquas quae eandem habent relationem ad res innumeras non quidem datas sed quae habent affectionem quandam communem ex re quadam data vel rebus quibusdam datis pendentem.

In a subsequent note he adds that it would be preferable to replace "invenire" with "datam vel datas esse ostendere".

In his Preface, Simson mentions the definition which Pappus recorded but denied being the true definition. It recurs in Simson's later translation of Pappus's account of the Porisms and there he adds a brief explanation in parentheses:[14]

A Porism is that which is lacking in hypothesis from a Local Theorem [that is to say, a Porism is a Local Theorem lacking or reduced in hypothesis].

Jones suggests that this definition has more meaning if "Local Theorem" is interpreted as a theorem about *position* rather than *locus*. Pappus goes on to say that according to the ancients a Porism is that by which something is proposed for discovering (investigating);[15] it is probably this definition which Simson has replaced.

Other definitions of Porism are found both from ancient and modern times.[16] We should note the definition given in 1794 by John Playfair, Professor of Mathematics at Edinburgh University:[17]

... a Porism may be defined, A proposition affirming the possibility of finding such conditions as will render a certain problem indeterminate, or capable of innumerable solutions.

Playfair argued that his definition is consistent with the way in which the Porisms must have been originally discovered and that with suitable interpretations it agrees with the definitions given by Pappus and Simson. Not surprisingly Trail disputed some of these claims. Playfair also noted that Simson's definition contained "a considerable degree of obscurity" and proposed the following "translation" of Simson's Latin as a remedy:

---

[14] "Porisma est quod deficit hypothesi a Theoremate Locali [hoc est, Porisma est Theorema Locale deficiens sive diminuta in hypothesi ejus]."
[15] "Porisma vero esse quo aliquid propositum est investigandum."
[16] See Jones's *Pappus* [17, pp. 548–553].
[17] "On the origin and investigation of porisms" [22]. Playfair's definition has been adopted by the *Oxford English Dictionary*. He noted that Dugald Stewart, son of Matthew Stewart, had given a similar definition a few years previously.

Dugald Stewart (1753–1828) was Professor of Mathematics at the University of Edinburgh during the period 1775-1785. From 1772 he had been assisting his father whose health was failing. In 1785 there was a remarkable switching of posts: Adam Ferguson (1723–1816) resigned from the Chair of Moral Philosophy and was given the Chair of Mathematics, apparently to provide him with an income; Dugald Stewart moved to Moral Philosophy and John Playfair (1748–1819) was appointed joint Professor of Mathematics with Ferguson. Playfair continued this process by moving to the Chair of Natural Philosophy in 1805; he is perhaps best remembered for his Parallel Postulate and his later involvement in Geology.

A Porism is a proposition, in which it is proposed to demonstrate, that one or more things are given, between which and every one of innumerable other things, not given, but assumed according to a given law, a certain relation, described in the proposition, is to be shown to take place.

We can only speculate on whether Simson (or Playfair) has really captured the true nature of Euclid's Porisms in his definition. Be that as it may, Simson proceeds to illustrate his definition by means of six elementary propositions.

Note on Proposition 1 (p. 18). The statement in the enunciation that the straight line AB is given in position means that an infinite straight line has been fixed and A and B are arbitrarily chosen points on it. Concerning the reference to the *Plane Loci* (p. 19) see the Introduction, p. 4.

Note on Proposition 2 (p. 20). Here we are concerned with the *circle of Apollonius* determined by $\frac{AF}{EF}$ = constant; E is the inverse of A with respect to the circle. Playfair also discussed this Porism [22, p. 168]. The lemma from the *Plane Loci* which Simson applies (p. 21) is the following:

Let ABC be a triangle, and let the straight line AD be drawn (to BC) so that BD is to DC as the square of BA is to the square of AC; I say that the rectangle BD, DC is equal to the square of AD.

Note on Proposition 4 (p. 23). (n) Simson's assertion in the composition that the rectangle AC, CB along with the square of DB is equal to the square of DC follows from

$$AC.BC = (DC + AD)(DC - DB) = (DC + DB)(DC - DB) = DC^2 - DB^2.$$

Note on Proposition 5 (p. 24). The details of the numerical example given at the end (p. 24) are as follows. If AB = 24 and BC = 12, then F = AH = AB + 2BC = 24 + 24 = 48 and since AH.BE = BC^2 we get BE = 144/48 = 3. Now if AD = 10 we have DE = AB − AD + BE = 24 − 10 + 3 = 17, so that F.DE = 48 × 17 = 816, AD.DB = 10 × (AB − AD) = 10 × 14 = 140 and DC = AB − AD + BC = 24 − 10 + 12 = 26, making DC^2 = 676. Thus AD.DB + DC^2 = 816 = F.DE as required by the Proposition.

Note on Fermat's Porisms (p. 31). Concerning Fermat's fifth Porism see Proposition 85 (p. 218) and its note (p. 245).

Note on Pappus's Account of the Porisms (p. 33). It should be borne in mind that what is presented here is the editor's English translation of Simson's Latin translation from the Greek.[18] Items in square brackets or italics are Simson's additions. (Cf. Jones's *Pappus* [17, pp. 94–104].)

---

[18]Simson's version was described by Trail ([40, Appendix III]) as "the Doctor's very improved translation of the general description of Euclid's Porisms".

The propositions stated in the second and third paragraphs are what have been referred to earlier as Pappus's two general propositions. The first of these has become known as the *Hyptios Porism* from the Greek ὕπτιος (see Simson's footnote (a) on p. 34);[19] it forms Simson's Proposition 19. The condition that "none of these intersections form an orbit",[20] which Simson has added to the general version of the second proposition, means that there is no closed polygon whose sides lie on the initial system of intersecting straight lines and whose vertices belong to the set of intersections which are required to lie on straight lines given in position. The triangular numbers, which also appear in the second result, are the numbers 1, $1 + 2 = 3$, $1 + 2 + 3 = 6$, $1 + 2 + 3 + 4 = 10$, ... with sides 1, 2, 3, 4, ...; in general the triangular number of side n is the number of items in the triangular array with 1 in the top level, 2 in the second level, 3 in the third level, ... and n in the nth (and final) level. Simson discusses the second proposition and the significance of orbits and triangular numbers in Proposition 21.

In Pappus's list of results from the three books of Porisms only the first has a detailed and unmutilated enunciation. It is discussed in Proposition 23. Simson also attempted to restore some of the Porisms from Pappus's brief statements; in particular, in Propositions 50, 53 and 57–62 he discusses at length the last three Porisms from Book 3.

---

[19] Jones disagrees with Simson's interpretation, arguing that παρύπτιος should apply to the case where two sides are parallel [17, pp. 392–393]. In fact, in a long letter to Dr Jurin dated 10 January 1724 [47] Simson explained the reasoning behind his translations of Pappus's terms and asked him to obtain Halley's opinion on the matter; I am not aware of any reply to this request.

[20] "Si nullae harum intersectionum in orbem redeant."

Part II. Pappus's Two General Propositions and Euclid's First Porism

This consists of Simson's Propositions 7–25 and includes his conjectures for the ten cases of Pappus's first general Proposition (Propositions 7, 8, 10–16, 19), the second general Proposition (Proposition 21), Euclid's first Porism (Proposition 23) and various applications and extensions of these results. Pappus's Lemmas 1–9 for the Porisms are developed in the course of this Part (see Appendix 3).

It is clear that all these propositions with the exception of the first are completely maimed and incomplete. However with a great deal of work we have elucidated a few, from which one may appreciate the excellence of the Porisms and how very greatly their loss is to be mourned by Geometers.

The ten Loci which were contained in the first of the preceding general Propositions seem to have been these which follow.

Locus 1: Proposition 7

"Suppose that there are four straight lines, two of which are parallel to each other, and that on one of these two points of intersection are given, while all but one of the remaining points of intersection lie on parallel straight lines which are given in position; this one will also lie on a straight line which is given in position."

Case 1. When the remaining two of the four straight lines are also parallel to each other, that is to say, when the four straight lines bound a parallelogram, and two points of intersection are given on one of the sides while one of the remaining points lies on a straight line which is given in position, then the other will also lie on a straight line which is given in position.

Let ABCD be a parallelogram, and let the points A, B be given on one of the sides while one of the remaining points C lies on the straight line EF, which is given in position; the other point D will also lie on a straight line which is given in position.

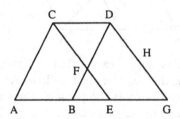

For since the straight line AB is given in magnitude, CD will also be given; and since the angle FEA is given, the angle FCD, which is equal to it, will be given. The straight line CD is therefore drawn with given magnitude and at the given angle FCD from the point C on the straight line EF, which is given in position, and consequently the point D lies on a straight line which is given in position, namely it is parallel to EF [*Dat.* 37].

Therefore let EG be positioned from the point E equal to AB, so that the points B, G are on the same sides of the points A, E; and let GH be drawn parallel to EF; then GH will be the straight line on which the point D lies; that is to say, if any parallel straight lines are drawn through A, B, the first of which meets the straight line EF in C, while the other meets the straight line GH in D, then the join CD will be parallel to AB. For triangles EAC, GBD are equiangular, therefore GB is to BD as EA is to AC; moreover

EA is equal to GB, because AB is equal to EG, therefore AC is equal to the straight line BD; and they are parallel; therefore CD is parallel to AB.

Or if EG is made equal to AB, as was said, and any parallelogram ABDC is drawn above AB with its vertex C lying on the straight line EF, then the join GD will be parallel to EF. For CD is equal to (AB, that is, to the straight line) EG, and they are parallel; therefore EC, GD are parallel.

Case 2. When two of the four straight lines are parallel, but the other two are not, and there are given on one of the parallel straight lines two points of intersection through which pass the two parallel straight lines given in position on which two of the remaining points lie, then the third point will also lie on a straight line which is given in position.

Let there be four straight lines AB, CD, AD, BC, of which AB, CD are parallel; and let the points A, B be given, while of the remaining points C, D, E one point C lies on a straight line AF which is given in position and another D lies on a straight line BG which is given in position and is parallel to the straight line AF; the remaining point E will also lie on a straight line which is given in position.

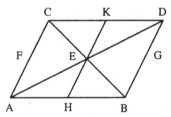

For since ABDC is a parallelogram, AE will be equal to ED; therefore since AD, drawn between the parallel straight lines AF, BG which are given in position, is cut in a given ratio at the point E, the straight line EH, which is drawn through E parallel to AF, BG, is given in position [*Dat.* 40].

Therefore let AB be bisected at H; the straight line HK, which is parallel to the straight line AF, will be the straight line on which the point E lies; that is to say, if through any point E on HK straight lines AD, BC are drawn to the straight lines BD, AC, then the join CD will be parallel to the straight line AB. For AE is to ED and CE is to EB as AH is to HB [VI 2]; therefore AE, ED are equal, as also are CE, EB. And since AE, EB are equal to DE, EC, respectively, and they contain equal angles, the angle ABC will be equal to the angle DCB [I 4]. Therefore CD is parallel to AB.

Or if CD is drawn parallel and equal to AB, and AD, BC are joined meeting each other in E, then the join EH will be parallel to BD. For AE is equal to ED since ABDC is a parallelogram, and AH is equal to HB; therefore EH is parallel to the straight line BD [VI 2].

Case 3. The rest remaining as in Case 2, suppose that both of the straight lines which are given in position do not pass through given points.

Let AB, CD, AC, BD be four straight lines of which AB, CD are parallel, and on one of them AB let two points of intersection A, B be given and moreover of the remaining three C, D, E let any two C, E lie on straight lines which are given in position and are parallel to each other, viz. let C lie on the straight line FC, and E on the straight line GE; the remaining point D will also lie on a straight line which is given in position.

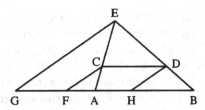

For since AF is to FG as AC is to CE, that is, as BD is to DE, and the ratio AF to FG is given, for the straight lines themselves are given, the ratio BD to DE will be given. Therefore since BE is drawn from the given point B to the straight line GE, which is given in position, and BE is cut in a given ratio at the point D, the straight line DH, which is drawn through the point D parallel to the straight line EG, will be given in position [*Dat.* 39].

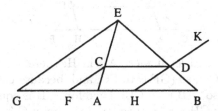

Therefore let GB be cut at H so that BH is to HG as AF is to FG, and let HK be drawn through H parallel to GE; then HK will be the straight line on which the point D lies. That is to say, if to some point E on the straight line GE there are drawn AE, BE of which AE meets FC in C, while BE meets HK in D, then the join CD will be parallel to AB. For AC is to CE as (AF is to FG, that is, by construction, as BH is to HG, that is, on account of parallel straight lines, as) BD is to DE. Therefore CD is parallel to AB.

Or if any straight line CD is drawn parallel to AB meeting FC, HK in C, D, and if AC is joined and meets GE in E, then the points B, D, E will be in a straight line. For, by construction, BH is to HG as AF is to FG, and, *componendo*, BG is to GH as AG is to GF; therefore the remaining part AB is to the remaining part FH or CD as (AG is to GF, that is, as) AE is to EC. Therefore the points B, D, E are in a straight line.

Or if to any point E on the straight line GE there are drawn AE, BE, of which AE meets FC in C, and CD is drawn to BE parallel to AB, then the join DH will be parallel to GE, FC. For BD is to DE as (AC is to CE, that is, as AF is to FG, that is, by construction, as) BH is to HG; therefore DH is parallel to the straight line GE.

Locus 2:   Proposition 8

"Suppose that there are four straight lines, two of which are parallel to each other, and that on one of these two points of intersection are given, while two of the remaining ones lie on straight lines which are given in position but are not parallel to each other; then the third point will also lie on a straight line which is given in position."

Case 1. When one of the straight lines which are given in position is parallel to the straight line on which there are two given points.

Let there be four straight lines AB, CD, AC, BD, of which AB, CD are parallel to each other, and let the points A, B be given; further, of the remaining points C, D, E let one of them C lie on the straight line GF which is given in position, and let another E lie on the straight line EF which is given in position and is parallel to the straight line AB; then the third point D will also lie on a straight line which is given in position.

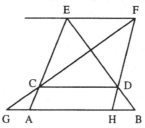

Let FD be joined and let it meet AB in H; therefore since GH, CD are parallel, as GH is to CD, so (GF is to FC, and, because GA, EF are parallel, so AE is to EC, and so) AB is to CD. Therefore GH is equal to AB; but AB is given, consequently GH is also given; and the point G is given, therefore H is given; and the point F is given, for GF, EF are given in position; therefore the straight line HF on which the point D lies is given in position.

Therefore let GH be put equal to the straight line AB, in such a way that the points B, H are on the same side of the points A, G, and let HF be joined; then HF will be the straight line on which the point D lies; that is to say, if to any point E on the straight line EF straight lines AE, BE are drawn which meet GF, HF in C, D, then the join CD will be parallel to the straight line AB. For since GH, AB are equal, GA will be equal to HB; but because GB, EF are parallel, as AC is to CE so (GA is to EF, that is HB is to EF, and so) BD is to DE. Therefore CD is parallel to AB. And the remaining converses of this Locus can be demonstrated more briefly.

Case 2. When two of the straight lines which are given in position pass through points which are given on one of the parallel straight lines.

Let there be four straight lines AB, CD, AC, BD of which AB, CD are parallel to each other, and let the points A, B be given; further, of the remaining points C, D, E let one C lie on the straight line BF which is given in position, and let another D lie on the straight line AF which is given in position; then the point E will also lie on a straight line which is given in position.

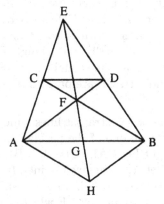

Let EF be joined and let it meet the straight line AB in G; and since CD is parallel to AB, triangles ACD, BCD will be equal [I 37] and consequently also triangles AFC, BFD. But triangle AFC is to triangle AFE as (the straight line AC is to the straight line AE, that is as BD is to BE, that is as) triangle BFD is to triangle BFE. Therefore triangle AFE is equal to triangle BFE. Moreover as triangle AFE is to triangle AFG so (the straight line EF is to the straight line FG, and so) triangle BFE is to triangle BFG. Therefore triangle AFG is equal to triangle BFG, and so the base AG is equal to the base GB. But AB is given, therefore the point G is given; and the point F is given, therefore the straight line GF on which the point E lies is given in position.

Therefore let AB be bisected at G, and let GF be joined; this will be the straight line on which the point E lies; that is to say, if to any point E on it straight lines AE, BE are drawn which meet the straight lines BF, AF in the points C, D, then the join CD will be parallel to AB. For since AG, GB are equal, the triangles AEG, BEG as also the triangles AFG, BFG will be equal, and so the triangles AEF, BEF are also equal. Moreover triangle AED is to triangle ADB as triangle FED is to triangle FBD, for both ratios are the same as the ratio ED to DB; and consequently triangle AEF is to triangle AFB as triangle AED is to triangle ADB [V 19], that is as the straight line ED is to the straight line DB. It will be shown in the same way that triangle BFE is to triangle BFA as the straight line EC is to the straight line CA. Since triangles AEF, BFE have in fact been shown to be equal, triangle BFE will be to triangle AFB as triangle AEF is to triangle AFB. Therefore ED is to DB as EC is to CA, and the straight lines CD, AB will be parallel [VI 2].

A shorter demonstration of this is found in Proposition 132 of Pappus's Book 7, which is his Lemma 6 for the Porisms, viz.

Let GH be put equal to GF, and let AH, HB be joined; then AHBF will be a parallelogram [I 4 and I 29], and consequently as AE is to EC so (HE is to EF, and) BE is to ED. Therefore CD is parallel to AB.

Another converse of this Locus is the following:

If any straight line CD is drawn parallel to the straight line AB, meeting the straight lines BF, AF, which are given in position, in C, D, and if the joins AC, BD meet each other in E, then there will be a straight line which passes through G, F, E. For since AB, CD are parallel, triangle AEF will be equal to triangle BEF, as was shown in the demonstration of the Locus. Moreover AE is to EC as (triangle AFE is to triangle EFC, that is as triangle BFE is to triangle EFC, that is as the straight line BF is to the straight line FC, or as) HA is to FC. And AH, CF are parallel, therefore the points G, F, E are in a straight line.

Case 3. When only one of the straight lines which are given in position passes through a given point on one of the parallel straight lines.

Let there be four straight lines AB, CD, AC, BD of which AB, CD are parallel to each other, and let A, B be given points; further, of the remaining points C, D, E let one D lie on the straight line AF which is given in position, and let another E lie on the straight line GF which is given in position; then the point C will also lie on a straight line which is given in position.

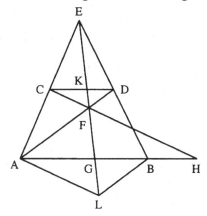

Let the join CF meet AB in H, and let GF meet the straight line CD in K. And since AB, CD are parallel, AG is to CK as GB is to KD; and CK is to GH as KD is to AG; therefore *ex aequali* AG is to GH as BG is to GA. But BG, GA are given, and so the ratio AG to GH is given; and AG is given, therefore GH is given, and the point H will be given. Moreover AF, GF are given in position, and so the point F is given; therefore the straight line FH on which the point C lies is given in position.

Therefore let it be arranged that AG is to GH as BG is to GA, and let HF be joined; this will be the straight line on which the point C lies; that is to say, if to any point E on the straight line GF straight lines AE, BE are drawn which meet HF, AF in C, D, then the join CD will be parallel to AB. This is proved in Proposition 133 of Pappus's Book 7 as follows, certain things having been supplied, as Commandino did in his notes:

Let the straight line AL be drawn parallel to CH to meet FG, and let LB be joined. Therefore since as BG is to GA so (AG is to GH, and, on account of the parallel straight lines AL, CH, so) LG is to GF, the triangles BGL, AGF are equiangular; therefore the angle FAG is equal to the angle LBG, and consequently LB is parallel to AD. Therefore, on account of the parallel straight lines AL, CH, as AC is to CE so (LF is to FE, and on account of the parallel straight lines LB, AD so) BD is to DE, and consequently CD is parallel to AB.

And if the point H is found in the way described, and AE, BE are drawn to any point E on the straight line GF so that BE meets AF in D, and DC is drawn parallel to AB to meet AE, then the points C, F, H will be in a straight line.

For since CD, AB are parallel, DK is to KC as (BG is to GA, that is as) AG is to GH; and *permutando* KD is to AG as CK is to GH. Moreover KF is to FG as (KD is to AG, that is as) CK is to GH; therefore triangles CKF, HGF are equiangular; consequently the points C, F, H are in a straight line.

Case 4. When neither of the two straight lines which are given in position passes through a given point on one of the parallel straight lines, and neither of them is parallel to the straight line on which the two points are given.

Let AB, CD, AC, BD be four straight lines of which AB, CD are parallel to each other, and let the points A, B be given; further, let two of the remaining points C, D, E lie on straight lines which are given in position; then the remaining point will also lie on a straight line which is given in position.

Article 1. Let AB, CD, AC, BD be four straight lines as has been stated, and let the points C, D lie on straight lines FC, GD which are given in position and are parallel to each other; the remaining point E will also lie on a straight line which is given in position.

Let EH be drawn parallel to DG; and since CFDG is a parallelogram, CD will be equal to FG; moreover AB is to CD, that is to FG as (BE is to ED, that is as) BH is to HG, and AB, FG are given, therefore the ratio BH to HG is given; and BG is given, therefore the point H is given, and the straight line HE is parallel to GD which is given in position, so that the straight line HE on which the point E lies is given in position.

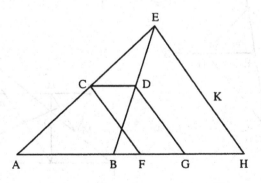

Therefore let it be arranged that BH is to HG as AB is to FG, and let the straight line HK be drawn through H parallel to the straight line GD; this will be the straight line on which the point E lies; that is to say, if to any point E on HK straight lines AE, BE are drawn which meet the straight lines FC, GD in the points C, D, then the join CD will be parallel to the straight line AB. For BH is to HG as BE is to ED, and since it has been arranged that BH is to HG as AB is to FG, by [V 12] or [V 19] AH will be to HF as BH is to HG (n); therefore AE is to EC as (BH is to HG, that is as) BE is to ED, and consequently CD is parallel to the straight line AB.

, And with the same construction, namely if it is arranged that BH is to HG as AB is to FG, and any straight line CD is drawn parallel to the straight line AB to meet the straight lines FC, GD which are given in position in C, D, and AC, BD meet each other in E, then the join EH will be parallel to the straight line GD, or FC. For BE is to ED as (AB is to CD, that is to FG, and, by construction, as) BH is to HG. Therefore EH is parallel to the straight line DG.

Article 2. Let the straight lines which are given in position not be parallel to each other: let AB, CD, AC, BD be four straight lines, of which AB, CD are parallel to each other, and let the points A, B be given; further, of the remaining points C, D, E let the first two C, D, which are on the straight line parallel to AB, lie on straight lines HF, GF which are given in position; then the remaining point E also lies on a straight line which is given in position.

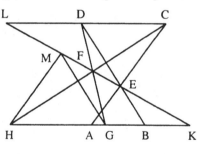

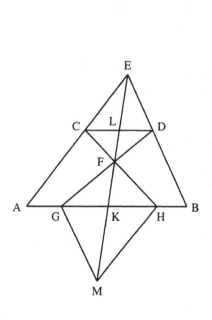

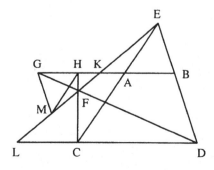

Let EF be joined and let it meet AB, CD in K, L; therefore since AK is to KB as (CL is to LD, that is as) HK is to KG, *componendo* or *dividendo*, or *dividendo inverse*, and *permutando*, AB will be to GH as BK is to KG; but AB, GH are given, and so the ratio BK to KG is given; and GB is given, therefore the point K is given; and F is given, therefore the straight line KF on which the point E lies is given in position.

Therefore let it be arranged that BK is to KG as AB is to GH, and let KF be joined; this will be the straight line on which the point E lies; that is to say, if AE, BE are drawn to any point E on the straight line KF and if they meet HF, GF in C, D, then the join CD will be parallel to AB. For let the straight line GM be drawn parallel to BE to meet EK, and let MH be joined. Therefore since BE is to GM as (BK is to KG, that is, by construction, as) AB is to GH, *permutando*, MG will be to GH as BE is to BA, and they are about equal angles which are alternate or one of them is interior and the other is exterior and opposite to it, therefore triangle ABE is equiangular with triangle HGM, and consequently angle BAE is equal to angle GHM, and so MH is parallel to the straight line AE. Moreover, on account of the parallel straight lines GM, EB, as GF is to FD so (MF is to FE, that is, because MH, AE are parallel, so) HF is to FC; and, *permutando*, GF is to FH as FD is to FC; therefore CD is parallel to the straight line GH, or AB [VI 2].

And, with the same construction, namely that by which it was arranged that BK is to KG as AB is to GH, if any straight line CD is drawn parallel to AB to meet the straight lines HF, GF, which are given in position, in C, D, and if the joins AC, BD meet each other in E, then the points E, F, K will be in a straight line.

This is Proposition 128 of Pappus's Book 7, which he demonstrates as follows.

Through G let the straight line GM be drawn parallel to the straight line DB, and let the join EF be produced to M. Therefore since BK is to KG as AB is to GH, and moreover BE is to GM as AB is to GH, because two straight lines are parallel to two straight lines, [a] therefore BE is to GM as BK is to KG; and BE is parallel to GM, therefore there is a straight line which

---

[a] The long explanation which Commandino adduces in his note B on p. 240. *a et seq.* for explaining this proposition is erroneous and inappropriate. And in note C at the words "quod duae duabus sint parallelae" [*translation*: because two straight lines are parallel to two straight lines] he expresses concern that these things have been added by somebody; but they are entirely necessary, for AK is parallel to CD by hypothesis, and GM is parallel to DB by construction, and it follows from the first of these that BE is to ED as AB is to CD, and from the other that CD is to HG as (DF is to FG, that is as) DE is to GM; therefore, *ex aequali*, AB is to HG as BE is to GM.

N.B. At note B in the text of Commandino "et permutando" must be put after "AB ad BK, ita HG ad GK" [*translation*: HG is to GK as AB is to BK].

passes through E, K, M, and the point F is on the straight line EM, therefore the points E, F, K are also in a straight line.

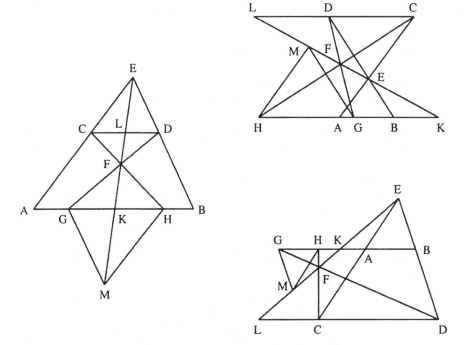

And the same things can be likewise shown in Case 3 when the points C, D lie on straight lines which are given in position.

Now in Article 2 of Case 4 let the points D, E, of which E is not on the straight line CD which is parallel to the straight line AB, be those points which lie on straight lines which are given in position, namely let D lie on the straight line GF and E on the straight line KF; then C will also lie on a straight line which is given in position.

For let CF be joined and let it meet AB in H, and let EF meet AB, CD in K, L, and since AB, CD are parallel AK will be to CL as KB is to LD; now CL is to KH as LD is to GK; therefore, *ex aequali*, AK is to KH as KB is to GK. Moreover KB, GK are given, and so the ratio AK to KH is given; and AK is given, therefore KH is given, and the point H will be given, and the point F is given because GF, KF are given in position, therefore the straight line HF on which the point C lies is given in position.

Therefore let it be arranged that AK is to KH as BK is to KG, and let HF be joined; this will be the straight line on which the point C lies; that is to say, if to any point E on the straight line KF straight lines AE, BE are drawn which meet the straight lines HF, GF in C, D, the join CD will be parallel to the straight line AB. For let the straight line GM be drawn parallel to the straight line BE to meet FK, and let MH be joined. And since AK is to KH as BK is to KG, by [V 12] or [V 19] AB will be to GH as (BK is to KG, and on account of the parallel straight lines BE, GM, as) BE is to GM, and *permutando* HG is to GM as AB is to BE; and they are about equal angles ABE, HGM, therefore triangles ABE, HGM are equiangular, and consequently angle BAE is equal to angle GHM, therefore MH is parallel to AE. Moreover GF is to FD as (MF is to FE, because GM, BE are parallel, and so) HF is to FC; therefore CD is parallel to AB [VI 2].

And if the point H is found in the manner stated and AE, BE are drawn to any point E on the straight line KF so that BE meets GF in D and DC is drawn parallel to AB to meet AE, then the points C, F, H will be in a straight line.

For since CD, AB are parallel, DL is to LC as (BK is to KA, that is, by construction, as) GK is to KH, and, *permutando*, DL is to GK as LC is to KH. But LF is to FK as (DL is to GK, that is as) LC is to KH; therefore triangles CLF, HKF are equiangular, and the points C, F, H will be in a straight line.

Case 5. When the point in which the two straight lines which are given in position meet is on the straight line on which two points are given; the rest remains as in Case 4.

Let there be four straight lines AB, CD, AC, BD of which AB, CD are parallel to each other and let the points A, B be given and of the remaining points C, D, E let one C lie on the straight line FC which is given in position, and let another E lie on the straight line FE which is given in position, and let the point F be on the straight line AB; then the point D will also lie on a straight line which is given in position.

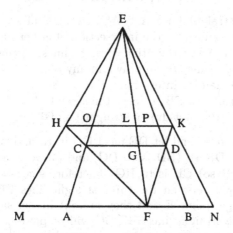

Let DF be joined, and let EF meet CD in G. Therefore since FC, FG are given in position, and CG is parallel to AB which is given in position, the triangle FCG is given in type; therefore the ratio FG to GC is given, and the ratio CG to GD is given, namely it is the same as the given ratio AF to FB, therefore the ratio FG to GD is given [*Dat.* 9], and the angle FGD is given; therefore triangle FGD is given in type [*Dat.* 44], and angle GFD will be given. Moreover FG is given in position, and the point F is given; therefore the straight line FD on which the point D lies is given in position [*Dat.* 32].

Therefore let any straight line HL be drawn parallel to the straight line AB between the straight lines FC, FG, and let HL to the fourth (proportional) LK be made as AF to FB; the join FK will be the straight line on which the point D lies; that is to say, if straight lines AE, BE are drawn from the points A, B to any point E on the straight line FG and meet FH, FK in C, D, then the join CD will be parallel to AB. For let the joins EH, EK meet the straight line AB in M, N; and let AE, FE, BE meet HK in the points O, L, P. And since MF is to FN as (HL is to LK, that is, by construction, as) AF is to FB, then MA will be to BN as AF is to FB [V 19]; and on account of the parallel straight lines HO is to PK as (MA is to BN, that is as) AF is to FB; therefore, *permutando*, PK is to FB as HO is to AF, and consequently KD is to DF as HC is to CF. Therefore CD is parallel to HK, and therefore also to AB. And this is Proposition 135 of Pappus's Book 7 and his Lemma 9 for the Porisms.

And with the same construction if any straight line AE is drawn which meets FH, FG in C, E, and CG is drawn through C parallel to AB, and the join BE meets the straight line CG in D, then there will be a straight line which passes through F, D, K. For FL is to LH as FG is to GC; now, by construction, as HL is to LK so (AF is to FB, and so) CG is to GD; therefore, *ex aequali*, FG is to GD as FL is to LK. Therefore the points F, D, K are in a straight line.

Proposition 9

"Let there be four straight lines AB, AC, BC, DE of which BC, DE are parallel to each other, and from the intersection of either one of the parallel straight lines DE and one of the other straight lines, for instance from the point D, let the straight line DF be drawn in any manner to AC, and let FB be joined; from the remaining intersection E of the same parallel straight line let EG be drawn parallel to FB to meet the straight line AB; then the join GC will be parallel to the first drawn straight line DF."

Let FD meet GE in K, and let DE meet GC in H; and since BF, EG are parallel, as FD is to DK so (BD is to DG, and on account of the parallel straight lines BC, DH so) CH is to HG. Therefore since two straight lines FDK, CHG are drawn between the three straight lines FEC, DEH, KEG which meet in a single point E and they are cut in the same ratio at the points D, H, then the straight lines FK, CG will be parallel to each other.

In another way.

Since GE, BF are parallel, EA is to AF as GA is to AB; and CA is to AE as BA is to AD; therefore *ex aequali in proportione perturbata* CA is to AF as GA is to AD, and *permutando*, DA is to AF as GA is to AC. Therefore triangles GAC, DAF are equiangular, so that angle CGA is equal to angle FDA, and consequently CG, FD are parallel.

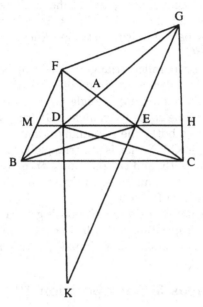

This is Proposition 134 of Pappus's Book 7 and his eighth Lemma for Euclid's Porisms; however we have stated it more fully. Pappus demonstrates the Proposition elegantly as follows.

Let BE, DC, FG be joined; triangle DBE will be equal to triangle DCE; let triangle DAE be added to both, therefore the whole of triangle ABE is equal to the whole of triangle CDA. Again since BF is parallel to EG, triangle BFE is equal to triangle BFG; let triangle ABF be taken away, and the remaining triangle ABE will therefore be equal to the remaining triangle AGF; but triangle ABE is equal to triangle ACD, therefore triangle ACD is also equal to triangle AGF; and when triangle ACG has been added to both, the whole triangle CDG will become equal to the whole triangle CFG; and it is on the same base CG; consequently CG is parallel to DF [I 39].

Corollary. And if the straight line DF is given in position, and the point C is given, the rest remaining as before, the point G of intersection of BD, EG will lie on the straight line CG which is given in position, for CG is drawn through the given point C parallel to the straight line DF which is given in position.

One of the converses of this Proposition is the following:

Suppose that from two points B, C two straight lines BF, CF are inflected, and again from the same points straight lines BG, CG are inflected; further, from the intersection F to either one of the second pair of inflected straight lines, for instance to BG which is drawn from the point B, let the straight line FD be drawn parallel to the other one CG of the second pair of inflected straight lines; and from the intersection G of the second pair of inflected straight lines to that one of the first pair of inflected straight lines which is drawn from the point C let the straight line GE be drawn parallel to the other one BF of the first pair of inflected straight lines; then the join DE will be parallel to the join BC.

For since FD, CG are parallel, triangle CDG will be equal to triangle CFG; let triangle CAG be taken away from both, and the remaining triangle CAD will be equal to the remaining triangle GAF. And since GE, BF are parallel, triangle BEF will be equal to triangle BGF, and when triangle BAF has been taken away from both, the remaining triangle BAE will be equal (to the remaining triangle GAF, that is) to triangle CAD; again let triangle DAE be taken away from both, and the remaining triangle BDE will be equal to the remaining triangle CED, and it is on the same base DE, therefore DE is parallel to the straight line BC [I 39].

This Proposition 9 is not contained in Pappus's general Proposition. However it seems that some Proposition with which it was connected, for instance the preceding Corollary, or some other similar result, was among Euclid's Porisms.

(p.91) ## Locus 3:   Proposition 10

"When two of four straight lines are parallel, and three points of intersection are given on one of the remaining straight lines while another point lies on a straight line which is given in position, then the remaining point will also lie on a straight line which is given in position."

Let there be four straight lines AB, AD, BD, CE of which BD, CE are parallel to each other, and on one of the remaining straight lines AB let three points A, B, C be given, while another D lies on a straight line FD which is given in position; then the remaining point E will also lie on a straight line which is given in position.

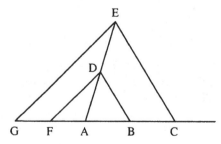

For since AD is to DE as AB is to BC, and AB, BC are given, the ratio AD to DE will be given. Therefore since AD is drawn from the given point A to the straight line FD which is given in position, and the point E is taken on AD making the ratio AD to DE given, the straight line EG which is drawn through E parallel to the straight line DF will be given in position [*Dat.* 39].

Therefore let it be arranged that AF is to FG as AB is to BC, and let GE be drawn parallel to FD; then GE will be the straight line on which the point E lies; that is to say, if any straight lines BD, CE are drawn parallel to each other from the points B, C to meet FD, GE in the points D, E, then the points A, D, E will be in a straight line. For since AF is to FG as AB is to BC, *invertendo*, CB will be to BA as GF is to FA, and *componendo* or *dividendo*, GA will be to AF as CA is to AB; therefore [V 12 or V 19] as CA is to AB so (GC is to FB (n), and so) CE is to BD, because triangles CEG, BDF are equiangular; therefore the points A, D, E are in a straight line.

And, with the same construction, if from the points B, C any two parallel straight lines BD, CE are drawn of which BD meets FD in D, and the join AD meets the straight line CE in E, then the straight line EG will be parallel to DF. For since on account of the parallel straight lines AD is to DE as (AB is to BC, that is, by construction, as) AF is to FG, then EG will be parallel to the straight line DF [VI 2].

Locus 4:   Proposition 11

"When no two of four straight lines are parallel, and three points of intersection are given on one of them, while two of the remaining points lie on straight lines which are given in position and are parallel to the one on which the given points lie, then the remaining point will also lie on a straight line which is given in position."

Let there be four straight lines AB, AD, BE, CE, and on one of them let three points of intersection A, B, C be given; of the remaining points let two D, E lie on straight lines GD, HE which are given in position and are parallel to AB; then the remaining point F will also lie on a straight line which is given in position.

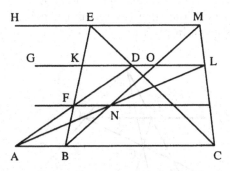

For since BE is drawn from the given point B to the parallel straight lines HE, GK which are given in position, the ratio BE to EK will be given [*Dat.* 38], therefore the ratio BC to KD is also given; but the straight line BC is given, and so KD is also given; and AB is given, therefore the ratio AB to KD is given. Moreover AB is to KD as AF is to FD; and consequently the ratio AF to FD is given. Therefore since GKD is given in position, the straight line which is drawn parallel to it through the point F will be given in position [*Dat.* 39].

Therefore let any straight line CLM be drawn from the point C, and let it meet the straight lines GK, HE which are given in position in L, M; further, let the joins AL, BM meet each other in the point N; the straight line drawn through N parallel to AB will be the straight line on which the point F lies; that is to say, if BE, CE are drawn to any point E on the straight line HE, and CE meets GK in D, while BE meets the straight line drawn through N parallel to AB in F, then the points A, F, D will be in a straight line. For let BM meet GK in O, and since as BC is to KD so (BE is to EK, and so BM is to MO, and so) BC is to OL, the straight line KD will be equal to the straight line OL. And BF is to FK as (BN is to NO, that is as) AB is to OL, or KD, therefore triangles ABF, DKF are equiangular, and consequently the points A, F, D are in a straight line.

Or when the point N has been found as described, if BE, CE are drawn to the straight line HE with CE meeting GK in D and the join AD meeting BE in F, then the join FN will be parallel to AB. For BF will be to FK as (AB is to KD, or OL, as was shown, that is as) BN is to NO; therefore FN is parallel to KO, that is to the straight line AB.

Locus 5:   Proposition 12

"When only one of the straight lines which are given in position is parallel to the straight line on which three points of intersection are given; the rest remains as in Locus 4."

Let there be four straight lines AB, AD, BE, CE, and on one of these let three points of intersection A, B, C be given; further, of the remaining points D, E, F let one D lie on the straight line GD which is given in position and is parallel to the straight line AB, and let another E lie on the straight line GH which is given in position; then the remaining point F will also lie on a straight line which is given in position.

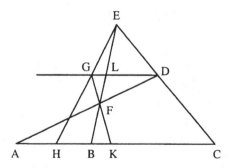

For let the join GF meet AB in K, and let BE meet GD in L; therefore since AB is to BK as (DL is to LG, that is as) CB is to BH, and the ratio CB to BH is given, the ratio AB to BK will be given; and AB is given, therefore

BK is given, and the point K is given; and the point G is given; therefore the straight line GK on which the point F lies is given in position.

Therefore let it be arranged that AB is to BK as CB is to BH, and let KG be joined; this will be the straight line on which the point F lies; that is to say, if any straight line CE is drawn to meet the straight lines GD, GH in D, E, and the join BE meets KG in F, then the points A, F, D will be in a straight line. For CB is to BH as DL is to LG, so that also AB is to BK as DL is to LG; and, *permutando*, AB is to DL as (BK is to GL, that is as) BF is to FL. And BFL is a straight line; therefore the points A, F, D are in a straight line.

Locus 6:   Proposition 13

91)

"When four straight lines meet each other, and three points of intersection are given on one of them, while two of the remaining points lie on straight lines which are given in position and pass through two of the given points."

Let there be four straight lines AB, AD, BE, CE, and let three points A, B, C be given on one of them, while of the remaining points D, E, F one D lies on the straight line BG which is given in position, and another E lies on the straight line AG which is given in position; then the remaining point F will also lie on a straight line which is given in position.

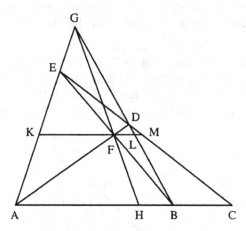

Let GF be joined, and let it meet AB in H; through F let a straight line be drawn parallel to the same straight line AB and let it meet AG, BG, CE in the points K, L, M. Therefore since as FM is to ML so AC is to CB and so KM is to MF, then the following will all be as FM is to ML: KM to MF, the remaining part KF to the remaining part FL (n), and AH to HB. Therefore AH is to HB as (KM is to MF, that is as) AC is to CB; and the ratio AC to CB is given, for the parts themselves are given; therefore the ratio AH to HB is given, and AB is given, thus the point H is given; and G is given; therefore the straight line GH on which the point F lies is given in position.

Therefore let it be arranged that AH is to HB as AC is to CB, and let HG
be joined; this will be the straight line on which the point F lies; that is to
say, if from the point C any straight line CDE is drawn to meet the straight
lines AG, BG in E, D, and the join BE meets HG in F, then the points A,
F, D will be in a straight line.

For since KM is to MF as (AC is to CB, that is as AH is to HB, that is as)
KF is to FL, the remaining part FM will be to the remaining part ML as (KM
is to MF, that is as) AC is to CB; and *permutando*, as FM is to AC so (ML is
to CB, and so) MD is to DC. Therefore the points A, F, D are in a straight
line. And this is Proposition 131 of Pappus's Book 7, and his 5th Lemma
for the Porisms. However it has been corrupted in no small measure in the
edition of Commandino, who has omitted the words *et reliqua ad reliquam*
[*translation:* and the remaining part to the remaining part] as if they were
superfluous and erroneous, but they are in no way superfluous, although they
are not quite correct; they are restored here to their unimpaired condition.
Also the figure is badly drawn there, for in it the points E, G, C (to which
E, F, B in our version correspond) are not in a straight line, as must be the
case.

Locus 7:   Proposition 14

"When only one of the straight lines which are given in position passes
through a given point; the rest remains as in the 6th Locus."

Let there be four straight lines AB, AD, BE, CE, and let the points A, B,
C be given and of the remaining points D, E, F let one D lie on the straight
line HDG which is given in position, while another E lies on the straight line
EHA which is given in position and passes through the given point A; then
the remaining point F will also lie on a straight line which is given in position.

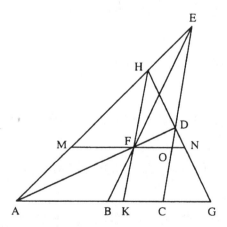

Let HF be joined, and let it meet AB in K; through F let MN be drawn
parallel to AB, and let it meet AH, GH, CE in M, N, O. Therefore since MF
is to FO as AB is to BC, while AC is to AG as FO is to FN, the ratio MF to

FN, that is the ratio AK to KG, will be the same as the ratio compounded of the ratios AB to BC and AC to AG, that is the ratio of the rectangle AB, AC to the rectangle BC, AG; but this ratio is given, because the points A, B, C, G are given. Therefore the ratio AK to KG is given, and AG is given, thus the point K is given; and H is given, therefore the straight line HK on which the point F lies is given in position.

Therefore let it be arranged that the straight line AK is to the straight line KG as the rectangle AB, AC is to the rectangle BC, AG, and let KH be joined; this will be the straight line on which the point F lies; that is to say, if any straight line CE is drawn to meet the straight lines GH, AH in D, E, while the join BE meets KH in F, then the points A, F, D will be in a straight line. For the ratio MF to FN (that is the ratio AK to KG) is the same as the ratio of the rectangle AB, AC to the rectangle BC, AG, that is the ratio compounded of the ratios AB to BC and AC to AG. Moreover the ratio MF to FN is the same as that compounded from the ratios MF to FO, that is AB to BC, and FO to FN; therefore when the common ratio AB to BC has been removed, the remaining ratio AC to AG will be the same as the remaining ratio FO to FN. Therefore *dividendo inverse*, and *permutando*, CG will be to ON, that is GD will be to DN, as AG is to FN. Therefore the points A, F, D are in a straight line.

91)

Locus 8:   Proposition 15

"When four straight lines meet each other, but none of them passes through a given point, and the straight lines which are given in position on which two of the points lie meet each other on the straight line on which three points of intersection are given."

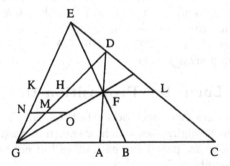

Let there be four straight lines AB, AD, BE, CE, and let three points A, B, C be given; further, of the remaining points D, E, F let two D, E lie on the straight lines GD, GE which are given in position and whose intersection G is on the straight line AB. The remaining point F will also lie on a straight line which is given in position.

For let GF be joined, and through F let KF be drawn parallel to AB, and let it meet the straight lines GD, GE, CE in the points H, K, L. Therefore since the ratio GB to BC is given, the ratio KF to FL will be given; and since the ratio CA to AG is given, the ratio LF to FH will also be given, and so (n1) the ratio KF to FH and also the ratio KF to KH are given [*Dat.* 9]. Moreover GH, GK are given in position, and KH is parallel to AB which is given in position; therefore triangle GKH is given in type [*Dat* 43], and so the ratio GK to KH is given, and consequently (n2) the ratio GK to KF is given [*Dat.* 9]; and the angle GKH is given, thus triangle GKF is given in type [*Dat.* 44], and the angle KGF will be given. Therefore, because GK is given in position, and the point G is given, the straight line GF is given in position.

Now since the ratio KF to FH is the same as (the ratio compounded of the ratios GB to BC and CA to AG, as was shown, that is as) the ratio of the rectangle GB, AC to the rectangle BC, GA [VI 23], *convertendo*, KF will be to KH as the rectangle GB, AC is to the rectangle AB, GC (n3). Therefore in order that the straight line GF may be found, let any straight line NM be drawn parallel to the straight line AB, meeting the straight lines GE, GD which are given in position in N, M, and let NO be taken on it which is to NM as the rectangle GB, AC is to the rectangle AB, GC; the join GO will be the straight line on which the point F lies. For let any straight line CD be drawn from the point C, and let it meet GM, GN which are given in position in D, E, and let the join BE meet GO in F; the points A, F, D will be in a straight line. Through F let the straight line KHFL be drawn parallel to the straight line NO or AB; therefore KF is to KH as (NO is to NM, that is as) the rectangle GB, AC is to the rectangle AB, GC; and, *convertendo*, the rectangle GB, AC is to the rectangle BC, GA as KF is to FH. Therefore the ratio KF to FH, that is the ratio compounded of the ratios KF to FL and FL to FH, is the same as the ratio compounded of the ratios GB to BC and CA to AG [VI 23], of which the ratio KF to FL is the same as the ratio GB to BC; therefore the remaining ratio LF to FH is the same as the remaining ratio CA to AG; and, *componendo* and *permutando*, as HL is to GC, that is as HD is to DG, so HF is to GA; and HF, GA are parallel, therefore the points A, F, D are in a straight line.

(p.92) ## Locus 9:   Proposition 16

"When four straight lines meet each other, but none of them passes through a given point, and the straight lines which are given in position on which two points of intersection lie are parallel to each other but not to the straight line on which the given points lie."

Let there be four straight lines ABC, ALK, BLH, CKH, and let there be three points of intersection A, B, C on one of them, while of the remaining points H, K, L let one H lie on the straight line FH which is given in position, and another K lie on the straight line EK which is given in position and is parallel to the straight line FH; then the remaining point L will also lie on a straight line which is given in position.

Through the point L let the straight line LM be drawn parallel to AB and let it meet the straight lines FH, EK, CH in the points M, N, O. And since ML is to LO as FB is to BC, and AC is to AE as LO is to LN, the ratio ML to LN (which is compounded of the ratios ML to LO and LO to LN,) will be the same as that which is compounded of the ratios FB to BC and AC to AE [V G], that is it will be the same as the ratio of the rectangle FB, AC to the rectangle BC, AE [VI 23]. But the ratio of the rectangles is given because their sides are given; therefore the ratio of the straight line ML to the straight line LN is given. Let LD be drawn to AB parallel to FH, EK, and since FD is to DE as ML is to LN, the ratio FD to DE will be given; and the straight line EF is given, therefore the point D is given, and so the straight line DL, which is parallel to a straight line which is given in position [*Dat.* 31], will be given in position. Therefore the straight line DL on which the point L lies is given in position.

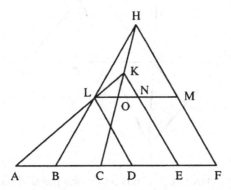

Therefore let it be arranged that the straight line FD is to the straight line DE as the rectangle FB, AC is to the rectangle BC, AE, and through D let DL be drawn parallel to the straight line EK; this will be the straight line on which the point L lies; that is to say, if any straight line AK is drawn to meet the straight lines DL, EK in L, K, while the joins BL, CK meet each other in H, then the join HF will be parallel to DL, EK.

For let LN be drawn parallel to the straight line AB, and let CH, EK, FH meet LN in the points O, N, M. It will therefore be shown as in the analysis that the ratio ML to LN is the same as the ratio (of the rectangle FB, AC to the rectangle BC, AE, that is, by construction, the ratio) of the straight line FD to the straight line DE. And LN is equal to DE, therefore LM is equal to DF, and they are parallel, thus the straight line FMH is parallel to the straight line DL.

For the tenth locus, Proposition 129 of Pappus's Book 7, which is his 3rd Lemma for the Porisms, is required, as also is Proposition 130, which is his 4th Lemma. These therefore have to be set forth first.

Proposition 17

(This is Proposition 129 of Pappus's Book 7.)

"If to three straight lines AB, CA, AD which come together in the same point two straight lines HE, HD are drawn; I say that as the rectangle which is contained by HE, GF to that contained by HG, FE, so is the rectangle HB, CD to HD, BC."

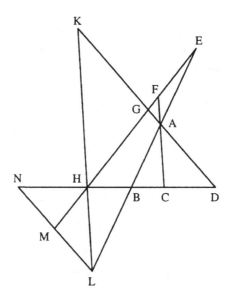

Through H let the straight line KL be drawn parallel to FC, and let DA, AB meet it in the points K, L; now through L let LM be drawn parallel to DA, and let it meet EH in M. And so since EH is to HL as EF is to FA, and moreover LH is to HM as AF is to FG, for in fact it is as HK to HG in the parallel scheme,† therefore, *ex aequali*, EH is to HM as EF is to FG; consequently the rectangle HE, GF is equal to the rectangle EF, HM. But the other rectangle is HG, FE; therefore as the rectangle HE, GF is to the rectangle HG, FE so the rectangle EF, HM is to the rectangle HG, FE, and so the straight line MH is to the straight line HG, and so LH is to HK. By the same argument‡ it is also shown that the rectangle HB, CD is to the rectangle HD, BC as LH is to HK. But we have shown that the rectangle

---

† Viz. AF is to FG as HK is to HG in the parallel scheme AFGHK, and LH is to HM as HK is to HG in the parallel scheme LHKGM. From this it is clear that Commandino did not understand what is meant by *parallelo* sc. schemate; for in the note to B in this Proposition he considers that the words of the Greek text have to be changed, while in fact they are correct.

‡ Commandino proves this at length and in a quite different way in note G in his commentary on this Proposition; but it is shown thus by a similar argument. Let LM meet BH in N; and since BC is to CA as BH is to HL, and

HE, GF is to the rectangle HG, FE as LH is to HK; therefore the rectangle HB, CD will be to the rectangle HD, BC as the rectangle HE, GF is to the rectangle HG, FE.

It will be shown by compound ratio in the following way.

Since the ratio of the rectangle HE, GF to the rectangle HG, FE is compounded of the ratio HE to EF and the ratio FG to GH, and HL is to FA as HE is to EF, and FA is to HK as FG is to GH, then the ratio of the rectangle HE, GF to the rectangle HG, FE will be compounded of the ratio HL to FA and the ratio FA to HK; but this compound ratio is the same as the ratio LH to HK. Therefore the rectangle HE, GF is to the rectangle HG, FE as the straight line LH is to the straight line HK.

By the same argument* it is also shown that the straight line LH is to the straight line HK as the rectangle HB, CD is to the rectangle HD, BC; therefore the rectangle HB, CD is to the rectangle HD, BC as the rectangle HE, GF is to the rectangle HG, FE.

Proposition 18

(This is Proposition 130 of Pappus's Book 7, but more fully enunciated.)

"If there are six points A, B, C, D, E, F in a straight line, and from three of them A, B, C three straight lines ALK, BLH, CKH are drawn meeting each other in the points L, K, H, and the joins LD, KE meet in G, and if the rectangle AF, DE is to the rectangle AD, EF as the rectangle AF, BC is to the rectangle AB, CF, then the points H, F, G will be in a straight line."

For since the rectangle AF, DE is to the rectangle AD, EF as the rectangle AF, BC is to the rectangle AB, CF, *permutando*, the rectangle AB, CF will be to the rectangle AD, EF as the rectangle AF, BC is to the rectangle AF, DE, that is as the straight line BC is to the straight line DE. But if KM is drawn through K parallel to AD and meeting BH in N, the ratio BC to DE will be composed of the ratio BC to KN, the ratio KN to KM, and finally the ratio KM to DE. Moreover the ratio of the rectangle AB, CF to the rectangle AD, EF is compounded of the ratio BA to AD and the ratio CF to FE; let the common ratio BA to AD be removed, this being in fact the same as the

---

LH is to HN as AC is to CD, for both ratios are the same as the ratio KH to HD in the parallel figures KHDAC, KHLND; *ex aequali*, BC will be to CD as BH is to HN; thus the rectangle HB, CD is equal to the rectangle NH, BC. But the other rectangle is HD, BC, therefore as the rectangle HB, CD is to the rectangle HD, BC so the rectangle NH, BC is to the rectangle HD, BC, and so the straight line NH is to the straight line HD, and so LH is to HK.

* For the rectangle HB, CD is to the rectangle HD, BC as the ratio compounded of the ratios HB to BC and CD to DH, that is the ratios LH to CA and CA to HK, which is the same as the ratio LH to HK.

ratio NK to KM, and the remaining ratio CF to FE will be the same as the ratio which is compounded of the ratio BC to KN, that is the ratio CH to HK, and the ratio KM to DE, that is the ratio KG to GE. Through E let EX be drawn parallel to CH, and let the join HG meet it in X, and the ratio CF to FE, which has been shown to be compounded of the ratio CH to HK and the ratio (KG to GE, that is the ratio) KH to EX, will be the same as the ratio CH to EX. Therefore since CF is to FE as CH is to EX, and CH is parallel to EX, there will be a straight line which passes through H, X, F, and consequently H, F, G will be in a straight line. *

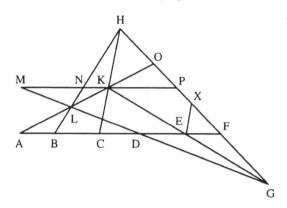

(p.93)   ## Locus 10:   Proposition 19

"When the two straight lines which are given in position are not parallel to each other, and they do not meet each other on the straight line on which the three points are given; the rest remains as in Locus 9. And this Locus is the most general of all."

---

* In the enunciation of the Proposition in the edition of Commandino there is, "Sit descripta figura ABCDEFGHKLMNX;" [*translation*: Suppose that the figure ABCDEFGHKLMNX has been drawn;] but he adds ineptly the last three letters MNX, which, he admits, are not found in the Greek codex; for they have nothing at all to do with the enunciation but refer to the construction which is necessary for the demonstration. Moreover we have changed the order of certain things at the end of the demonstration for the sake of clarity. Also in the text of Commandino on p. 243.*a* of Pappus's Book 7 in the fifth line from the end of the demonstration for CE read GE; and in the third and second lines from the same end after the words "CH ad EX" [*translation*: CH to EX] there should be read, igitur proportio CF ad EF eadem est quae CH ad EX; cum vero CX ipsi EX parallela sit, recta erit linea quae transit, &c. [*translation*: therefore the proportion CF to EF is the same proportion which CH has to EX; now since CX is parallel to EX, there will be a straight line which passes etc.]

Let there be four straight lines ABC, ALK, BLH, CKH, and let three points of intersection A, B, C be given on one of them, while of the remaining points H, K, L one L lies on the straight line GDL which is given in position and another K lies on the straight line GEK which is given in position; then the remaining point H will also lie on a straight line which is given in position.

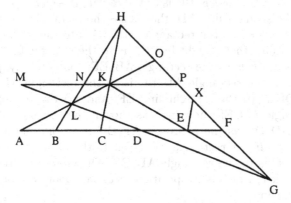

For let GH be joined and let it meet the straight line AB in F and the straight line AK in O. Therefore since the two straight lines AO, AF have been drawn to the three straight lines GL, GK, GO which come together at the same point G, by Proposition 129 of Pappus's Book 7, which is Proposition 17 of this book, the rectangle AF, DE will be to the rectangle AD, EF as the rectangle AL, KO is to the rectangle AO, LK. Again since the two straight lines AO, AF have been drawn to the three straight lines HB, HC, HF, by the same Proposition, the rectangle AF, BC will be to the rectangle AB, CF as the rectangle AL, KO is to the rectangle AO, KL. Therefore the rectangle AF, DE is to the rectangle AD, EF as the rectangle AF, BC is to the rectangle AB, CF, and, *permutando*, the rectangle AB, CF is to the rectangle AD, EF as the rectangle AF, BC is to the rectangle AF, DE, that is as the straight line BC is to the straight line DE. But the ratio of the straight line BC to the straight line DE is given, for the segments themselves are given, therefore the ratio of the rectangle AB, CF to the rectangle AD, EF is given; and the ratio of the side AB to the side AD is given, because the sides themselves are given, therefore the ratio of the remaining side CF to the remaining side EF is given [*Dat.* 65]; but the straight line CE is given, and so the point F is given, and the point G is given, therefore the straight line GF on which the point H lies is given.

Now to find the point F, the ratio which CF has to FE has to be found, which, as is clear from *Dat.* 65, is the same as the ratio which is compounded of the ratio AD to AB and the ratio BC to DE, and moreover this compound ratio is the same as the ratio of the rectangle AD, BC to the rectangle AB, DE. Therefore let it be arranged that the straight line CF is to the straight line FE as the rectangle AD, BC is to the rectangle AB, DE, and let the point F found in this way and the point G be joined; GF will be the straight line on which the point H lies; that is to say, if any straight line AL is drawn from the point A and it meets the straight lines GD, GE in the points L, K, and moreover the joins BL, CK meet each other in H, then the points H, F, G will be in a straight line, or, in other words, the point H will lie on the straight line GF.

For since the rectangle AF, BC is to the rectangle AD, BC as (the straight line AF is to the straight line AD, that is as) the rectangle AF, DE is to the rectangle AD, DE, and as the rectangle AD, BC is to the rectangle AB, DE, so, by construction, (the straight line CF is to the straight line EF, and so) the rectangle AB, CF is to the rectangle AB, EF; consequently, *permutando*, and by [VI 1], as the rectangle AD, BC is to the rectangle AB, CF so (the straight line DE is to the straight line EF, and so) the rectangle AD, DE is to the rectangle AD, EF; now, as was shown, the rectangle AF, DE is to the rectangle AD, DE as the rectangle AF, BC is to the rectangle AD, BC; therefore, *ex aequali*, the rectangle AF, DE is to the rectangle AD, EF as the rectangle AF, BC is to the rectangle AB, CF. Therefore, by Proposition 130 of Pappus's Book 7, which is Proposition 18 of this book, the points H, F, G are in a straight line.

The tenth Locus can in fact be established in a shorter way using the method by which we previously established it in the *Philosophical Transactions* for the year 1723, No. 377. This is as follows, the enunciation remaining as in the preceding discussion.

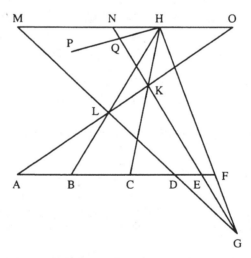

Through the point H let the straight line HM be drawn parallel to the straight line AB, and let it meet the straight lines GD, GE, AL in the points M, N, O. And since on account of the parallel straight lines MH is to HO as DB is to BA, while AC is to CE as OH is to HN, the ratio MH to HN, which is in fact compounded of the ratio MH to HO and the ratio OH to HN, will be the same as that which is compounded of the ratio DB to BA and the ratio AC to CE, that is the ratio of the rectangle AC, BD to the rectangle AB, CE. But the ratio of the rectangles is given because their sides are given; therefore the ratio of the straight line MH to the straight line HN is given. Let the join GH meet AB in F, and since MH is to HN as DF is to FE, the ratio DF to FE will be given, and the straight line DE is given, therefore the point F is given, and the point G is given; therefore the straight line GF on which the point F lies is given in position.

Therefore let it be arranged that the straight line DF is to the straight line FE as the rectangle AC, BD is to the rectangle AB, CE, and let GF be joined; this will be the straight line on which the point H lies; that is to say, if any straight line ALK is drawn which meets the straight lines GD, GE in L, K, and the joins BL, CK meet each other in H, then the points H, F, G will be in a straight line.

For when the same things have been constructed as in the analysis, as was shown in it, the ratio MH to HN will be the same as (the ratio of the rectangle AC, BD to the rectangle AB, CE, that is, by construction, as) the ratio of the straight line DF to the straight line FE. Therefore, *dividendo*, or *componendo*, the ratio MN to NH will be the same as the ratio DE to EF, and, *permutando*, MN will be to DE, that is NG will be to GE, as NH is to EF; and NH is parallel to the straight line EF, therefore the points H, F, G are in a straight line.

And if from the points B, C straight lines BH, CH are drawn in any manner to meet on the straight line GF which has been found, and if BH, CH meet GD, GE which are given in position in the points L, K, then the points A, L, K will be in a straight line.

For through the point H let MH be drawn parallel to the straight line AB, and let it meet the straight lines GL, GK in M, N and the join AK, produced if necessary, in O. Therefore on account of the parallel straight lines MH is to HN as DF is to FE; moreover the ratio MH to HN is compounded of the ratio MH to HO and the ratio OH to HN; and the ratio DF to FE, by construction, is compounded of the ratio DB to BA and the ratio AC to CE; therefore the ratio compounded of the ratio MH to HO and the ratio OH to HN is the same as the ratio compounded of the ratio DB to BA and the ratio AC to CE, of which the ratio OH to HN is the same as the ratio AC to CE, on account of the parallel straight lines; therefore the remaining ratio MH to HO is the same as the remaining ratio DB to BA. Therefore, *permutando*, as MH is to DB, that is as HL is to LB, so HO is to BA; and again, *permutando*, as HL is to HO so LB is to BA, and they are about equal alternate angles; therefore the triangles HLO, BLA are equiangular, and so the points A, L, O and consequently the points A, L, K are in a straight line.

Corollary 1. It is clear that the straight line FG which is the Locus of the points H in this Proposition passes through the point G, namely the point of intersection of the straight lines GD, GE which are given in position; and consequently if some other point H on the Locus is found, which will be done by drawing in any manner from the point A a straight line AL which meets the straight lines GD, GE which are given in position in L, K and by joining BL, CK which meet each other in H, then the join GH will be the straight line which is the Locus to be found. Moreover a similar result holds in the preceding material after Proposition 7 with the exception of Proposition 9.

Corollary 2. The following Problem is easily solved from these observations:

Three points A, B, C having been given on a straight line, and three straight lines GD, GE, PQ having been given in position, to locate the triangle among them whose sides will pass through the three given points.

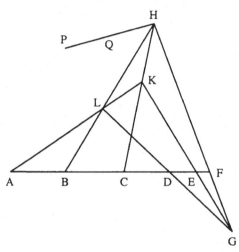

Suppose that it has been done, and let it be triangle HKL whose angle H is on the straight line PQ, angle K is on the straight line GE, and angle L is on the straight line GD, and whose sides KL, HL, HK pass through the given points A, B, C. And since there are four straight lines AB, ALK, BLH, CKH, and three points of intersection A, B, C are given on one of them, while of the remaining points H, K, L two K, L lie on the straight lines GE, GD which are given in position, then the remaining point will also lie on a straight line which is given in position by this Proposition. Therefore let this straight line, which is GF by Corollary 1 of this Proposition, be found and let PQ, the third straight line which is given in position, meet it in H, and let BH, CH be joined which meet GD, GE in L, K; triangle HKL will be the triangle which was to be found, for the straight line KL passes through the point A by the second part of this Proposition.

(p.93)   **Proposition 20**

(This demonstrates the use of Porisms in the solution of Problems.)

"Two straight lines AB, CD meeting each other in E having been given in position, to draw to them from a given point F two straight lines FCA, FBD such that the rectangles AE, EB, CE, ED contained by the segments cut off between the straight lines drawn and the point E are equal to given areas."

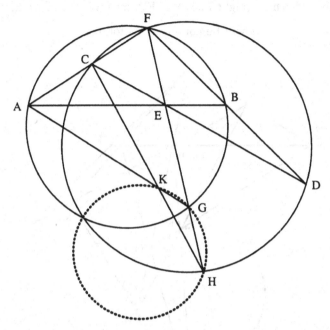

Suppose that it has been done. Let FE be joined, and moreover let the rectangle FE, EG be equal to the given rectangle AE, EB and the rectangle FE, EH be equal to the given rectangle CE, ED. Therefore since FE is given in position and magnitude, the straight lines EG, EH will be given in position and magnitude [*Dat.* 61], and because of the equal rectangles the points A, F, B, G will lie on a circle, and likewise the points C, F, D, H will lie on another circle [Converse of III 35]. Let the joins GA, HC meet in K; the angle FGA will be equal to the angle FBA in the same segment and likewise the angle FHC will be equal to the angle FDC, thus the angle GKH will be equal to the given angle BED [I 16]. And since the straight line GH is given in position and magnitude, the point K lies on the circumference of a circle which is given in position by Proposition 2, Book 1 of Apollonius's *Plane Loci*, which was published in Glasgow in 1749. Now the three straight lines FA, GA, HC are drawn from the three collinear points F, G, H and two of their points of intersection A, C lie on the straight lines EB, ED which are given in position; therefore the remaining point of intersection K will also lie on a straight line which is given in position by the preceding Proposition 19. But it has been shown that the same point K lies on the circumference of a circle which is given in position; therefore the point K is given, and the point G is given, and so the straight line GK is given in position; and the straight line AE is given in position, therefore the point A is given, and E is given, and consequently the straight line AE is given in magnitude. And the rectangle

AE, EB is given, therefore the straight line EB is given in magnitude [*Dat.* 61]; but it is also given in position; consequently the point B is given, and F is given, therefore the straight lines FA, FB are given in position.

It will be composed as follows.

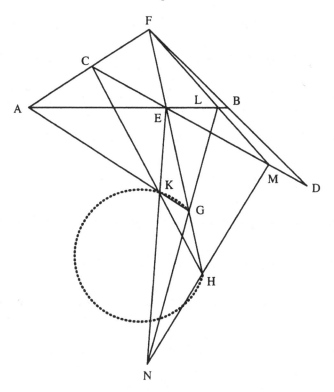

From the point F let a straight line FL be drawn in any manner and let it meet AB, CD which are given in position in the points L, M; let LG, MH be joined, meeting each other in N, and on GH let the segment of the circle be described which subtends an angle equal to the given angle BED, and let the join EN meet the circumference of this circle in K, while the join GK meets the straight line BE in A; finally let the rectangle AE, EB be made equal to the given rectangle FE, EG, and let FA, FB be joined; these will be the straight lines which had to be found, that is to say if FA, FB meet the straight line CD which is given in position in the points C, D, then the rectangle CE, ED will be equal to the given rectangle FE, EH. For by Corollary 1 of the preceding Proposition the straight line EN is the Locus which has to be found, and consequently the point K will lie on the straight line HC, for since the straight line FA meets the straight lines AB, CD which are given in position in the points A, C, the joins GA, HC will meet each other in a point of the straight line EN, and the straight lines GA, EN meet each other in the point K, which consequently will be on the straight line

HC. Now since the rectangle AE, EB is equal to the rectangle FE, EG, the points A, F, B, G will lie on a circle, and so the angles FBA, FGA will be equal, and, by construction, the angles BED, GKH are equal, therefore the angles BDC, or EDC, and GHK, or FHC, are equal; and consequently the points F, D, H, C lie on a circle. Therefore the rectangle CE, ED is equal to the given rectangle FE, EH.

Pappus's second general Proposition follows in which he extends what he enunciated in the first for just four straight lines to any number of straight lines.

94)

Proposition 21

"Suppose that any number of straight lines meet each other, there being not more than two through the same point; now let all the points on one of them be given, and moreover for each point on another of the straight lines let there be a straight line which is given in position on which it lies; or more generally thus: suppose that any number of straight lines meet each other and that there are not more than two through the same point, while all the points on one of them are given; the number of remaining points will be a triangular number, whose side expresses the number of points which lie on a straight line which is given in position; if no three of these intersections are located at the angles of a triangular area [* no four are at the angles of a quadrilateral, no five are at the angles of a pentagon, etc., i.e. generally, if none of these intersections form an orbit], then for each one of the remaining points of intersection there will be a straight line given in position on which it lies."

Explanation of the second Proposition.

It is to be noted that the number of intersections which are to be found on one straight line for any proposed quantity of straight lines, of which no more than two pass through the same point and none are mutually parallel, is one less than the number of straight lines; for two cut each other mutually in a single point, then if a third straight line is drawn it will cut the previous pair in two points, a fourth will cut the three previous straight lines in three points, etc.; therefore the number of intersections on three straight lines is one increased by two, i.e. three; the number of the same on four straight lines is three increased by three; then on five straight lines it is the last of the previous cases, or six, increased by four, etc.; as is clear, these numbers are triangular numbers, the side of each one of them being the number of intersections which are found on any one straight line, i.e. the number which is one less than the total number of straight lines. Therefore if the total

---

* From necessity we have added to Pappus's text those things which are enclosed in brackets, for without them the Proposition would not be true beyond the case of five straight lines.

number of given points, which is the same as the number of intersections on any one straight line, is taken away from this number of intersections, the remaining number will still be triangular, namely that whose side is one less than the side of the previous triangular number and consequently two less than the number of straight lines proposed. And this is the number of intersections which in this Proposition Pappus requires to lie on straight lines which are given in position, and if no three are at the angles of a triangle, no four at the angles of a quadrilateral, and so on, he asserts that each one of the remaining intersections lies individually on a straight line which is given in position.

In fact the Proposition divides conveniently into two cases, which are also quite clearly indicated by Pappus, who set forth the hypothesis of the simpler case in a Proposition which is elegant and has the greatest generality of this type.

Case 1

Let there be any number of straight lines, e.g. six, AF, BG, CH, DK, EL, EA and suppose that all the points on one of them are given, namely the points A, B, C, D, E, while all the points on another F, M, N, O lie on straight lines which are given in position; then each remaining intersection will lie on a straight line which is given in position.

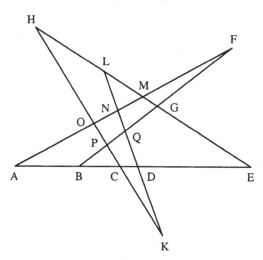

For let any one of the remaining intersections be taken, e.g. L; since there are four straight lines AE, AM, DL, EL and three points A, D, E are given on one of them, while of the remaining points L, M, N two M, N lie on straight lines which are given in position, by Proposition 19 the third point L will lie on a straight line which is given in position; the same is shown in the same way for all the remaining points.

Case 2

Now suppose that all the points lying on straight lines which are given in position do not lie on the same straight line (the number of such points is two less than the number of straight lines proposed) but that none of them form an orbit, the rest remaining as before; then all the remaining points of intersection will lie on straight lines which are given in position.

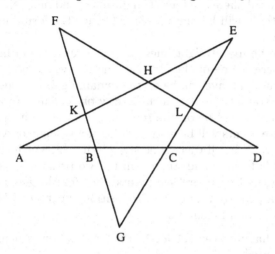

Lemma 1

Suppose that any number of straight lines meet each other and that there are no more than two through the same point; let any of the straight lines be taken and suppose that the number of intersections which is made up by taking two points on each of the chosen straight lines is equal to the number of these straight lines; then these points will form an orbit.

For since there are two points on each individual straight line, there will be at least three on two straight lines, and four on three, and so on, from which it is clear that the number of points will always be at least one more than the number of straight lines unless the last straight line passes through the first point; i.e. unless the points form an orbit, in which case alone will the number of points be equal to the number of straight lines.

Lemma 2

Suppose that any number of straight lines meet each other and that there are no more than two through the same point; now let any of the intersections be taken whose number is equal to the number of all the straight lines; then either all these intersections, or some of them, will form an orbit, that is to say they are located at the angles of a polygon or a triangle.

For three intersections of three straight lines are at the angles of a triangle; now if there are four straight lines, and four points are taken, one of these will necessarily be found on each individual straight line; but if on one of the four straight lines only one point is to be found, the remaining three will be on the remaining three straight lines, and consequently they will be at the angles of a triangle; if there is no straight line on which only one point is to be found, there will be two on each of the four straight lines; and there are four points, therefore, by Lemma 1, they are at the angles of a quadrilateral. And it is clear that if there are four straight lines, and more than four points are taken, then there will be more possibilities for the formation of an orbit by some of them.

Now let there be five straight lines, and let five points of intersection be taken, and, if there is one of the straight lines on which none of these five points is to be found, all five will be on the remaining four straight lines; now if there is some straight line on which exactly one point is to be found, the remaining four points will be on the remaining four straight lines; therefore in both cases some points will be at the angles of a triangle or a quadrilateral by the preceding case; but if there is no straight line on which either no point or a single point is to be found, there will be two points on each one of the five straight lines, and there are five points; therefore by Lemma 1 they are at the angles of a pentagon; and the Proposition is extended likewise to six straight lines, and so on indefinitely.

These things having been set forth, the Proposition is demonstrated in the following manner.

First part. Let there be five straight lines AD, AE, BF, CG, DH, and when the points given on one of the straight lines, namely A, B, C, D, have been removed, there will be six points E, F, G, H, K, L left on four straight lines; and let three of these (for the side of the triangular number 6 is 3) which are not at the angles of a triangle formed by three of the straight lines set forth, e.g. E, F, G, lie on straight lines which are given in position; then the remaining three H, K, L will also lie on straight lines which are given in position.

For since there are four straight lines AE, BF, CG, DF, and three points of intersection E, F, G are taken on them, there will be one of these straight lines on which necessarily only one of these points is to be found; for otherwise there are the following two possibilities: either there will be some straight line on which there is no point, and consequently there will be three points on the three remaining straight lines, that is at the angles of a triangle, contrary to hypothesis; or there will be at least two of these three points on each of the four straight lines, and consequently there would be at least four points; but there are only three; therefore there is necessarily some straight line on which only one point is to be found. Let this straight line be AE on which there is in fact the point E; therefore the remaining two F, G are on the three straight lines BF, CG, DF; therefore since there are four straight lines BD, BF, CG, DF, and three points of intersection B, C, D are given on one of the straight lines, while of the remaining points F, G, L two F, G lie on straight lines which are given in position, the remaining point L on those three straight lines lies on a straight line which is given in position by the preceding Proposition 19. Now let GE be taken, namely the straight line from these three which passes through the point E on the fourth straight line AE; and all points on this straight line GE will lie on straight lines which are given in position. Therefore by the first case of this Proposition the remaining points H, K lie on straight lines which are given in position.

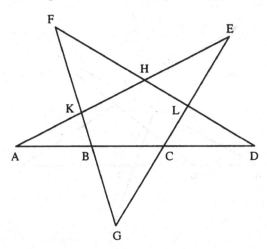

Second part. Now let there be six straight lines AE, AF, BG, CH, DK, EL; and when the five given points A, B, C, D, E which are on one of the straight lines have been removed, ten points F, G, H, K, L, M, N, O, P, Q will be left on five straight lines; and, by hypothesis, four of these (for the side of the triangular number 10 is 4) lie on straight lines which are given in position and none of them form an orbit; let these be F, G, H, K; then the remaining six L, M, N, O, P, Q will lie on straight lines which are given in position.

For since there are five straight lines AF, BG, CH, DK, EL, and four points of intersection F, G, H, K are taken on them, there will be one straight line on which only one of these points is to be found; for otherwise there are the following two possibilities: either there will be some straight line on which there is no point, and so the four points would be on four straight lines, and consequently by Lemma 2 some of them will form an orbit, contrary to hypothesis; or there will be at least two points on each straight line, and consequently there would be at least five points; but there are only four, and so there is necessarily some straight line on which only one point is to be found; let this straight line be AF, on which there is the point F; therefore the remaining three points G, H, K are on the remaining four straight lines BG, CH, DK, EL; therefore since there are five straight lines BE, BG, CH, DK, EL, and all the points B, C, D, E on one of the straight lines are given, and of the remaining points three G, H, K which are not at the angles of a triangle lie on straight lines which are given in position, by the first part of this demonstration, the remaining three points on these four straight lines, namely L, P, Q, lie on straight lines which are given in position. Now let BF be taken, namely the straight line which passes through the point F on the fifth straight line AF; and all points on this straight line BF will lie on straight lines which are given in position, thus by the first case of this Proposition the remaining points M, N, O lie on straight lines which are given in position.

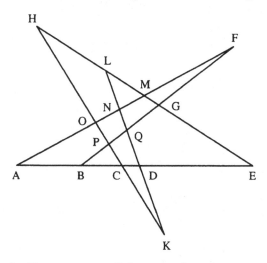

As is clear, the Proposition will be proved in the same straightforward manner for seven, eight, etc. straight lines, i.e. for any number of straight lines.

(p.95)   ## Proposition 22

(This is Proposition 127 of Pappus's Book 7,
and his first Lemma for the Porisms.)

"Let the figure ABCDEFG be drawn, let AD be to DC as AF is to FG, and let HK be joined; I say that HK is parallel to AC."

Or in more detail.

"Suppose that A, F, D, G, C are points on a straight line such that AF is to FG as AD is to DC; let FE, GE be inflected to the straight line AB, and let DB, CB be inflected to the same straight line; let the straight lines inflected from the points F, D meet each other in K, and let the straight lines inflected from the points G, C meet in H, and let HK be joined; then KH will be parallel to AC."

Through F let FL be drawn parallel to BD; therefore since AF is to FG as AD is to DC, *permutando* and by [V 12], CA will be to AG as AD is to

AF, or because of the parallel lines as BA is to AL; therefore the join LG is parallel to BC. Therefore because of the parallel lines EK is to KF and EH is to HG as EB is to BL. Consequently EH is to HG as EK is to KF, and so HK is parallel to AC.

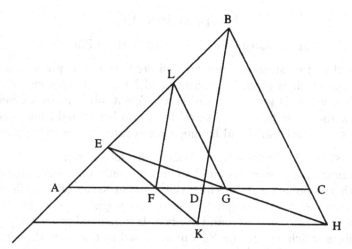

The proof which is found in maimed form in Commandino and, as it seems, in the Greek text may be restored in the following way using compound ratio.

Since AD is to DC as AF is to FG, *invertendo* CD will be to DA as GF is to FA; and *componendo, permutando* and *convertendo*, AC is to CG as AD is to DF.* But the ratio AD to DF is compounded of † the ratio AB to BE and the ratio EK to KF; and the ratio AC to CG is compounded of ‡ the ratio AB to BE and the ratio EH to HG (n). Therefore the ratio compounded of the

---

* It is given thus in Commandino, but it will be demonstrated a little more briefly as follows: Since AF is to FG as AD is to DC, *permutando*, DC will be to FG as AD is to AF, and by [V 12] AC is to AG as AD is to AF, therefore, *convertendo*, AC is to CG as AD is to DF.

† ‡ What is contained in these notes depends on the following Lemma, which is in fact given in the second part of Proposition 3 of Pappus's Book 8. See the figure for Proposition 22 here.

Suppose that two straight lines BA, AD meet two straight lines EK, KB, and let any one of them, say AD, be called the first, another AB the second, either of the remaining straight lines, say KB, the third and the other KE the fourth; the ratio of the segment of the first AD between the intersection of the first and the second A and the third to the segment DF of the same between the third and the fourth will be compounded of the ratio of the segment AB of the second between the intersection A and the third to the segment BE of the same between the third and the fourth and the ratio of the segment KE of the fourth between the third and the second to the segment KF of the same between the third and the first.

ratio AB to BE and the ratio EK to KF is the same as the ratio compounded of the ratio AB to BE and the ratio EH to HG. Let the common ratio AB to BE be removed; the remaining ratio EK to KF is the same as the ratio EH to HG. Therefore HK is parallel to AG.

(p.95) ## Proposition 23

(This is Porism 1 of Book 1 of Euclid's Porisms.)

"Suppose that two straight lines are inflected from two given points to a straight line which is given in position, and let one of them cut off from a straight line which is given in position a segment adjacent to a given point on it; then the other will also take off from another straight line a segment 'adjacent to a given point' and having a given ratio 'to the other segment'."

In No. 377 of the *Philosophical Transactions* for the year 1723 this Porism was explained as if it were required to find not only that other straight line along with the point on it to which the segment taken off is adjacent, but also the ratio which this segment has to the segment taken off from the first straight line. However I found afterwards innumerable mutually parallel straight lines which satisfy the Porism proposed in this way; therefore Pappus's words "and having a given ratio" must be interpreted as "having the same ratio as a given ratio", and so indeed they must be written to distinguish clearly those things which are given by hypothesis in the Porism from those which are asserted to be given and are proposed for finding. And in this sense of the Proposition, which is certainly the correct one, only two straight lines satisfy the Porism. The investigation of the Porism is now carried out as follows.

Let A, B be the given points from which straight lines AE, BE are inflected in any manner to the straight line CD which is given in position; moreover let one of them AE take off from the line CF which is given in position a segment GH adjacent to the given point H; it has to be shown that the other straight line BE takes off from some particular straight line a segment which is adjacent to some given point on it and which has the same ratio to the other segment GH as the given ratio which the given straight line β has to the given straight line α, and this other straight line has to be found and shown to be given, and the point on it to which the segment to be taken off is adjacent has to be shown to be given.

For let the straight line FL be drawn to AB parallel to KB; therefore the ratio AD to DF is the same as the ratio AB to BL which is compounded of the ratio AB to BE and the ratio (EB to BL, that is the ratio) EK to KF.

In note G of his commentary at this place, namely Proposition 127 of Pappus's Book 7, Commandino takes the join GL to be parallel to BC; but this is not to be assumed at all, for, if it is, almost the whole of the first proof is assumed, and if this were done, the second proof would be entirely useless.

Suppose that the Porism is true, that KL is the straight line from which the segment has to be taken off by the straight line from the point B and that M is the point to which the segment is adjacent. Therefore let BE meet KL in N, and NM will be the segment cut off. Therefore by hypothesis GH has the same ratio to NM as the straight line α has to the straight line β. Let the join AB meet CD in D, and let CK, KL meet AD in the points F, L. Therefore since AD, BD are drawn from the points A, B to the straight line CD, and AD takes off from the straight line CF the segment FH adjacent to the point H, while BD takes off from KL the segment LM adjacent to the point M, then FH will have to LM the same ratio as α to β. Therefore GH is to NM as HF is to LM, so that also GF is to LN as HF is to LM, or as α is to β.

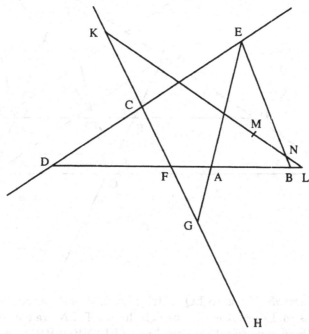

Again let AO, BO be inflected from the points A, B to the straight line CD and let them meet CF, KL in the points P, Q; it will be shown similarly that PF is to LQ as HF is to LM. And it has been shown that GF is to LN as HF is to LM; therefore PF is to LQ as GF is to LN. Let the straight line GR be drawn to EB parallel to AB, and let it meet the straight line CD in S, and let the join CR meet AB in T. And since DA is to AB as SG is to GR, that is as DF is to FT, and the ratio DA to AB is given, for the points D, A, B are given, the ratio DF to FT will be given. But DF is given, for CF is given in position, therefore FT is given, and the point T will be given and the straight line CT will be given in position, for the point C is given. Therefore CT, CF are given along with their ratio, and so the ratio RT to GF is also given. Let OB meet CR in V, and let PV be joined; and since DA is to AB as DF is to FT, and AO, BO are inflected to the straight line CD from the points A, B, while FC, TC, which meet AO, BO in P, V, are inflected to the same straight line CD from the points F, T, the straight line PV will be parallel to AB [Prop. 22], and so also to GR. Therefore VT is to PF as RT is to GF; now it has been shown that PF is to LQ as GF is to LN;

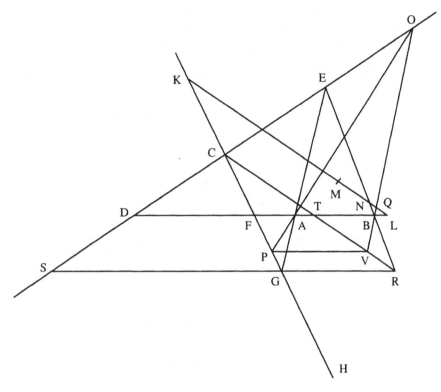

therefore, *ex aequali*, VT is to LQ as RT is to LN; and, *permutando*, NL is to LQ as RT is to TV; that is, the straight lines RT, LN are cut in the same ratio by the three concurrent straight lines TBL, VBQ, RBN; therefore RT is parallel to LN. [b] But it has been shown that the ratio which RT has to GF is given, as also is the ratio GF to LN, namely the same as the given ratio α to β; therefore the ratio RT to LN is given [*Dat.* 9]. Now RT is to LN as TB is to BL, so that the ratio TB to BL is given; and TB is given, because the points T, B are given, therefore BL is given, and so the point L is given. And since KL is drawn through the given point L parallel to the straight line CT which is given in position, KL is given in position [*Dat.* 31]. And since FH is given along with its ratio to LM, then LM will be given; and the point L is given, thus the point M is also given. Therefore the straight line KL and the point M are given in position. This is what had to be investigated.

---

[b] This is shown by the following Lemma.

It will be composed as follows.

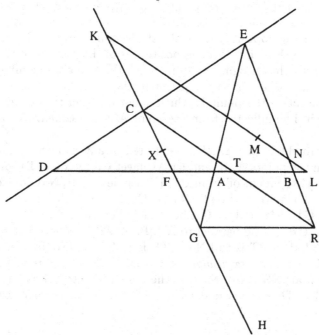

Let the straight line through the given points A, B meet CD, CF which are given in position in the points D, F; let CF to FX be made as α to β; and let DF to FT be made as DA to AB, and let the join CT be produced; also let TB to BL be made as CT to FX, the segment BL being located on the straight line BD on either side of B; and let LK be drawn parallel to CT.

"If between three concurrent straight lines AB, AC, AD are drawn two straight lines BCD, EFG which are cut in the same ratio at C, F, then the straight lines drawn will be parallel to each other."

For let the straight lines CH, FK be drawn to AB parallel to AD; therefore since, by hypothesis, EF is to FG as BC is to CD, because of the parallel lines BH will be to HA as EK is to KA; and, *componendo* and *permutando*, as BA is to EA so (HA is to AK, and so) CA is to AF. Therefore BC, EF are parallel.

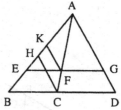

(Or as follows, the same construction being retained.) Since BH is to HA as (BC is to CD, that is, by hypothesis, as EF is to FG, that is as) EK is to KA, and KA is to KF as HA is to HC, therefore, *ex aequali*, BH will be to HC as EK is to KF. And they are about equal angles, therefore triangles BHC, EKF are equiangular, and the angle CBH will be equal to the angle FEK, and consequently BD, EG are parallel.

Further let H be the given point on CF to which the segment cut off by the straight line inflected from the point A is adjacent, and let FH to LM be made as CF to FX, the segment LM being located on the straight line LK on the same or the opposite side of the straight line DL to FH, according as the point L is on the same or the opposite side of the point B to the point T. Then KL will be the straight line from which the straight line inflected from the point B cuts off a segment adjacent to the point M, whose ratio to the segment adjacent to the point H which is cut off from the straight line CF by the straight line inflected from the point A is the same as β to α, or FX to FC.

For let straight lines AE, BE be inflected in any manner to the straight line CD from the points A, B, and let AE meet CF in G and BE meet KL in N; moreover let them cut off from CF, KL segments GH, NM adjacent to the points H, M. Then GH will be to NM as CF is to FX, that is as α to β. For let BE meet CT in R, and let GR be joined. Therefore since DF is to FT as DA is to AB, GR will be parallel to AB [Prop. 22]; therefore FC is to CT as GF is to RT; but as RT is to LN so (TB is to BL, and so, by construction,) TC is to FX; therefore, *ex aequali*, as GF is to LN so (CF is to FX, and so it was made that) FH is to LM. And therefore GF is to LN as GH is to NM [V 12 or V 19]. Thus the segment cut off GH is to the segment cut off NM as α is to β.

(p.95)   ## Proposition 24

(This is a Lemma for the Problem in the following Proposition; it is moreover a Case of Locus 1 of Book 2 of Apollonius's *The Cutting off of a Ratio*.)

"To draw to two straight lines AB, AC which are given in position from a given point D outside them, a straight line DFE which will cut off from the straight lines given in position segments EB, FC which are equal to each other and are adjacent to the given points B, C."

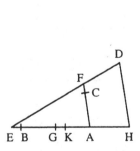
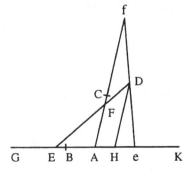

Suppose that it has been done; now let DH be drawn to AB parallel to AC, and further let BG be put equal to AC; therefore BG is given, and the point G is given. Also let HK be put equal to HD, and since EA is to AF as

EH is to HD, that is (because EG is equal to AF, and KH is equal to HD) EH is to HK as AE is to EG, also, *componendo* or *dividendo*, EK will be to KH as AG is to GE. Therefore the rectangle GE, EK is equal to the given rectangle KH, GA; but the straight line KG is given, because the points K, G are given, consequently GE is given [*Dat.* 85 or 86], and the point E will be given; and the point D is given, therefore the straight line DE is given in position.

Therefore let the rectangle GE, EK be made equal to the given rectangle KH, GA [VI 28 or VI 29] (n), and let DE be joined, meeting AC in F, and let HK be made equal to HD, and let BG be made equal to AC; the segments EB, FC will be equal to each other. For since, by construction, EK is to KH as AG is to GE, then, *componendo* or *dividendo*, AE will be to EG as (EH is to HK, or HD, that is as) EA is to AF. Therefore EG is equal to AF, and BG is equal to the straight line AC, consequently the remaining part EB is equal to the remaining part FC.

Proposition 25

(This provides an example by which the use of Porisms
in the solution of Problems is demonstrated.)

"To inflect from two given points A, B to the straight line CD which is given in position straight lines AE, BE which will cut off from two other straight lines CF, YZ which are given in position segments GH, ZY adjacent to the given points H, Y, which will have the same ratio among themselves as the given ratio α to β."

Suppose that it has been done; and since AE, BE have been inflected from the given points A, B to the straight line CD which is given in position, one of them AE cutting off the segment GH adjacent to the given point H from CF which is given in position, by the Porism in Proposition 23 the other BE will cut off from a certain other straight line a segment adjacent to a given point to which the other GH will have the same ratio as the given ratio α to β. Let this straight line be found, let it be KL, let M be the point found on it, and let BE meet KL in N. Therefore GH is to NM as α is to β; now, by hypothesis, the same GH is to ZY as α is to β; therefore NM is equal to ZY. Therefore since from the given point B the straight line BN has been drawn cutting off from KL, ZY which are given in position segments NM, ZY adjacent to the given points M, Y, which are equal to each other, BN will be given in position [Prop. 24]; therefore the point E in which BN meets the straight line CD which is given in position is given; consequently AE, BE are also given in position.

The composition is clear, namely when the straight line KL and the point M have been found by means of the Porism in Proposition 23, in addition let the straight line BNZ which cuts off equal segments NM, ZY adjacent to the given points M, Y from the straight lines KL, ZY be found by Proposition

24, and let BZ meet the straight line CD in E, and let AE be joined, meeting
CF in G; HG will be to YZ as α is to β. For by construction GH is to NM
as α is to β, and, likewise by construction, NM is equal to ZY; therefore GH
is to ZY as α is to β.

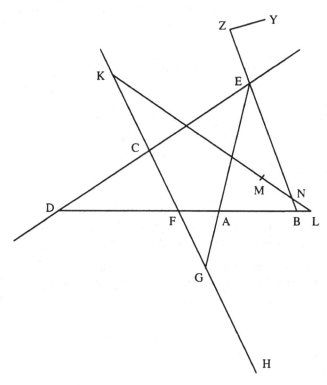

Notes on Part II

Note on Proposition 8 (p. 50). In Case 2 we have Simson's first specific reference to Pappus's Lemmas for the Porisms, which were intended as an aid to readers of Euclid's three Books. In fact Simson discusses all of Pappus's Lemmas in the course of his *Treatise*. See Appendix 3 and Jones's *Pappus* [17, pp. 260–295].

(n) In Case 4, Article 1, the third paragraph, from $\frac{BH}{GH} = \frac{AB}{FG}$ we get $\frac{AB}{BH} = \frac{FG}{GH}$, so that $\frac{AB+BH}{BH} = \frac{FG+GH}{GH}$, i.e. $\frac{AH}{BH} = \frac{FH}{GH}$, from which it follows that $\frac{AH}{FH} = \frac{BH}{GH}$ as claimed.

Note that the possibilities that the points which lie on lines which are given in position are C, D in Case 3 and D, E in Case 4 are mentioned after the discussion of Pappus's Proposition 128 in Case 4. Simson ignores the possibility in Case 5 that C, D are the points which lie on lines which are given in position, but his argument easily adapts to this.

In the penultimate paragraph of Case 5 LK is defined as a "fourth proportional". Simson pointed out in his edition of the *Elements* that what purported to be Euclid's demonstration of Proposition 18 of Book 5 is faulty in that it assumes without proof the existence of such quantites (see Simson's note on V 18, [29, pp. 365–368] (1756 Latin edition) or [30, pp. 381–385] (1756 English edition) and T.L. Heath [15, Vol. II, p. 170]). In characteristic fashion he asserted that this demonstration could not have been that given by Euclid, but must have been substituted by a later commentator in the belief that he had found a shorter demonstration.

Note on Proposition 10 (p. 60). (n) In the last sentence of the third paragraph of the demonstration from $\frac{GA}{FA} = \frac{AC}{AB}$ we get $\frac{AC}{AB} = \frac{GA+AC}{FA+AB} = \frac{GC}{FB}$ as claimed.

Note on Proposition 13 (p. 63). (n) In the second sentence of the second paragraph of the demonstration from $\frac{KM}{FM} = \frac{FM}{LM}$ we get $\frac{FM}{LM} = \frac{KM-FM}{FM-LM} = \frac{KF}{FL}$ as claimed.

Note on Proposition 15 (p. 65). The condition in the enunciation "but none of them passes through a given point" would seem to apply to AD, BE, CE and is included to ensure that the figure can vary.

(n1) In the second paragraph of the demonstration from $\frac{KF}{FL}$ and $\frac{FL}{HF}$ given we get that $\frac{KF}{HF} = \frac{KF}{FL} \cdot \frac{FL}{HF}$ and $\frac{KF}{KH} = \frac{KF}{KF-HF}$ are both given; then (n2) from $\frac{GK}{KH}$ given we deduce that $\frac{GK}{KF} = \frac{GK}{KH} \cdot \frac{KH}{KF}$ is given.

(n3) In the first sentence of the next paragraph from $\frac{KF}{HF} = \frac{GB.AC}{BC.GA}$ we get $\frac{KF}{KH} = \frac{KF}{KF-HF} = \frac{GB.AC}{GB.AC-BC.GA} = \frac{GB.AC}{(GC-BC).AC-BC.(GC-AC)} = \frac{GB.AC}{GC.AC-BC.GC} = \frac{GB.AC}{AB.GC}$ as claimed.

Note on Proposition 16 (p. 66). As in Proposition 15 a condition has apparently been included to ensure that the figure can vary – here it applies to ALK, BLH, CKH.

(n) Simson's reference to [V G] identifies Proposition G which he annexed to Book 5 in his edition of Euclid's *Elements*. Its statement is as follows [30, p. 174]:

> If several ratios be the same to several ratios, each to each; the ratio which is compounded of ratios which are the same to the first ratios, each to each, shall be the same to the ratio which is compounded of ratios which are the same to the other ratios, each to each.

Note on Proposition 17 (p. 68). It is important for later applications to realise that the rectangles in the two ratios are related by projection through A: E ↔ B, F ↔ C, G ↔ D, H ↔ H. In fact the Proposition asserts the invariance of the cross ratio {HF, EG}.

Simson made substantial annotations in a 1589 edition of Commandino's *Pappus*.[21] In one of his notes, which is dated 2 September 1759, he recorded the following shorter demonstration of Pappus's Proposition 129.

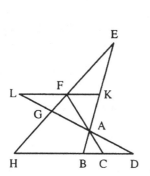
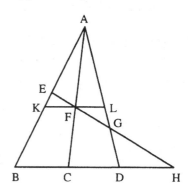

Through the point F let FK be drawn parallel to HD, and let it meet BE, DA in K, L and since the ratio of the rectangle HG, FE to the rectangle HE, GF is compounded of the ratio HG to GF, and FE to EH, the same ratio will be compounded of the ratio HD to LF and the ratio FK to HB, that is of the ratios HD to DC, DC to LF and the ratio FK to HB. But the ratio DC to LF is the same as the ratio BC to KF; consequently the ratio of the rectangle HG, FE to the rectangle HE, GF is compounded of the ratios HD to DC, BC to FK and FK to HB, that is of the ratios HD to DC, and BC to HB. But the ratio of the rectangle HD, BC to the rectangle HB, DC is compounded of the same. Therefore the ratio of the rectangle HG, FE to the rectangle HE, GF is the same as the ratio of the rectangle HD, BC to the rectangle HB, CD and

---

[21]These are preserved in Glasgow University Library. See [45] and [44] (a transcription of the original annotations).

invertendo the rectangle HB, CD is to the rectangle HD, BC as the rectangle HE, GF is to the rectangle HG, FE. Q.E.D.

Simson also noted that Pappus's Proposition 142 of Book 7 is the converse of his Proposition 129 and that the above proof is the converse of his proof of the former.

Note on Proposition 19 (p. 70). This is the Hyptios Porism, the first of Pappus's general propositions which Simson restored in his *Philosophical Transactions* paper [26]. His second demonstration is essentially the same as that given in the paper – there are small changes in notation, terminology and diagrams (see Appendix 1). It appears that Newton had already elucidated the Hyptios Porism – a statement of it along with some of its special cases is contained in a manuscript of Newton's dating from the 1690s.[22]

Simson's comment at the end of Corollary 1 presumably applies to all the Loci; Proposition 17 should be ruled out as well as Proposition 9.

Note on Proposition 20 (p. 74). Here we are concerned with two elementary properties of circles or the converses of these properties:
(i) if RS, TU are chords of a circle which intersect in V, then RV.VS = TV.VU;
(ii) the angle subtended by a chord is the same for all points on the circumference which lie on the same side of the chord.

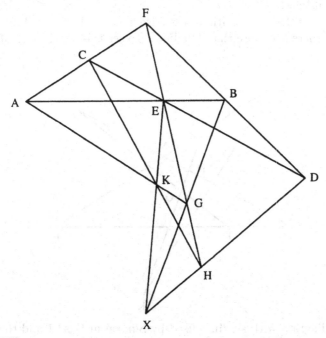

[22] See D.T. Whiteside [42, Vol. 7, pp. 263–269] and also Jones's *Pappus* [17, pp. 570–572].

Simson claims to apply Proposition 19 to the system of lines FH, FA, GA, HC, for which the intersections F, G, H are given and of the remaining three A lies on AE, which is given in position, and C lies on CE, which is given in position. Since AE, CE meet on the straight line on which the three given points F, G, H lie, it seems that Proposition 15 would be a more precise reference. In either case the locus of K is a straight line through E.

Simson apparently ignores the fact that this straight line will meet the given circle through G and H in two points, i.e. there are two possible positions for K. The second position is in fact the intersection of DH with BG. To see this apply Proposition 19/15 to the system of lines FH, FB, GB, HD, which is just another position of the previous system (A → B, C → D), and so X, the intersection of HD and GB, must lie on the same straight line through E as K. Further X also lies on the given circle through G and H since ∠GXH = ∠DHF − ∠XGH = ∠DHF − ∠BGF = ∠DCF − ∠BAF = ∠AEC = ∠BED = ∠GKH. However, if we choose X rather than K, the analysis continues as before with A and B interchanged.

Note on Proposition 21 (p. 77). Apart from minor textual variations Simson's account here is the same as that given in his *Philosophical Transactions* paper [26] (see Appendix 1). For "orbit" and "triangular numbers" see the note on Pappus's account of the Porisms, pp. 42–43. With reference to the first paragraph of Lemma 1 note that the intersections have to be made by the chosen lines.

Simson notes that Proposition 21 is false in general without his orbit condition when there are more than five lines. To see this consider the following diagram:

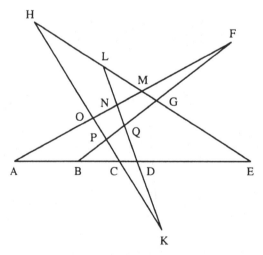

This is the diagram with six lines used by Simson in Case 1 and the Second Part of Case 2 in which A, B, C, D, E are the given collinear points. Let O, N, M, F vary on lines which are given in position. Then by Case 1 all the other points H, K, P, Q, G, L vary on lines which are given in position. Now choose from these lines the ones on which F, M, H, P vary. These points form an orbit in Simson's terminology but no three of them do. Choose F arbitrarily on its line. Then the parts of the figure shown below are determined:

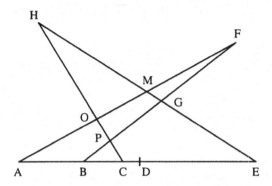

However, the line KDQNL is not determined and so K, Q, N, L are not restricted to lie on lines which are given in position.

In the case of five lines three points of intersection are required to lie on lines which are given in position; in this case Pappus's condition coincides with Simson's orbit condition.

Note on Proposition 22 (p. 82). (n) Concerning the point which Simson marked with ‡ note that $\frac{AC}{CG} = \frac{AB}{BL} = \frac{AB}{BE}\cdot\frac{BE}{BL} = \frac{AB}{BE}\cdot\frac{HE}{HG}$ as claimed. Pappus's Lemma which is quoted in the footnote for † and ‡ is what we usually call Menelaus's theorem.

Note on Proposition 23 (p. 84). Simson's earlier interpretation and treatment of this Porism will be found in Appendix 1, where I have also included a note on the two versions.

Newton discussed this Porism in an earlier unpublished manuscript.[23]

Note on Proposition 24 (p. 88). (n) The point E is determined by the condition that GE.EK is equal to a given quantity, G and K being given. In general this will give two positions for E (E, e in Simson's second diagram), so that the required line DFE will not be unique.

The order of events in the final paragraph, the composition, has to be changed: G, H, K have to be constructed first as in the first part of the proof; then GE.EK is set equal to KH.GA.

Note on Proposition 25 (p. 89). Again there is non-uniqueness, although this is not clear from Simson's discussion or terminology. Recall that both Propositions 23 and 24, which are applied here, do not produce unique lines.

---

[23]See Whiteside [42, Vol. 7, pp. 242–245] and Jones's *Pappus* [17, pp. 570–572].

Part III. Lemmas and Restorations

Propositions 26–79 contain the rest of Pappus's Lemmas for the Porisms (10–38) (see Appendix 3) and Simson's restorations of several of Euclid's Porisms, notably the last three of Book 3 as recorded by Pappus (Propositions 50, 53, 57, 58, 61, 62).

Eight Lemmas which are contained in Pappus's Book 7, namely in Proposition 136 and the seven subsequent Propositions, are useful for the following Porism, and although the Porism can be established without them, in order that their use should no longer escape the notice of Geometers, it did not seem to be a useless enterprise to present those Lemmas here.

Proposition 26

(This is Proposition 136 of Pappus's Book 7 and his tenth Lemma for the Porisms; it is moreover the converse of Proposition 17 of the present work.)

"Let two straight lines HD, HE be drawn from the same point to two straight lines BAE, DAG, and let points C, F be taken on HD, HE; and let the rectangle contained by EF, GH be to that contained by EH, FG as the rectangle which is contained by DH, CB is to that contained by DC, BH. I say that there is a straight line which passes through the points C, A, F."

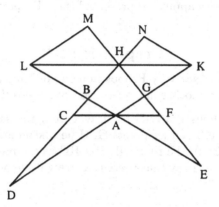

Let KL be drawn through H parallel to CA to meet AB, AD in the points L, K; and through L let LM be drawn parallel to AD, and let EH be produced to M. Finally through K let KN be drawn parallel to AB, and let DH be produced to N. And so since on account of the parallel lines DC is to CB as DH is to HN,* the rectangle DH, CB will be equal to the rectangle DC, HN. But the other specified rectangle is that which is contained by DC, BH; therefore as the rectangle DH, CB is to the rectangle DC, BH so the rectangle DC, HN is to the rectangle DC, BH, that is, so the straight line HN is to the straight line BH. But it is assumed in fact that the rectangle EF, GH is to the rectangle EH, FG as the rectangle DH, CB is to the rectangle DC, BH; moreover as the straight line HN is to the straight line BH, so, because of the parallels, the straight line KH is to the straight line HL, and so the straight line GH is to the straight line HM, and so the rectangle EF, GH is

---

* For DH is to HK as DC is to CA; then CA is to CB as HK is to HN, because of the similar triangles KHN, ACB; therefore, *ex aequali*, DC is to CB as DH is to HN.

to the rectangle EF, HM. Therefore the rectangle EF, GH is to the rectangle EF, HM as the rectangle EF, GH is to the rectangle EH, FG; therefore the rectangle EH, FG is equal to the rectangle EF, HM. Consequently GF is to FE as MH is to HE; *componendo* and *permutando*, HE is to EF as ME is to EG. But LE is to EA as ME is to EG; therefore HE is to EF as LE is to EA, and consequently AF is parallel to LHK. But CA is also parallel to the same LK; therefore there is a straight line which passes through C, A, F. This is what it was required to demonstrate.

It should be noted that Pappus's Proposition 142 in Book 7 is the same as this Proposition 136, although he gives another demonstration for it. But both demonstrations apply to the figures of both Propositions, as will be clear from the following demonstration, which is the one Pappus uses in Proposition 142. Moreover the lettering is as it is found in Proposition 142. It is not easy to guess the reason why Pappus has presented these Propositions as if they were different.

(p.188) <h1 style="text-align:center">Proposition 27</h1>

(This is that very Proposition 142 of Pappus and it
contains another demonstration of Proposition 136.)

"Let two straight lines DB, DE be drawn from the same point D to two straight lines AB, AC, and let points G, H be taken on DE, DB; and let the rectangle BH, CD be to the rectangle BD, HC as the rectangle EG, FD is to the rectangle ED, GF. I say that there is a straight line which passes through the points A, G, H."

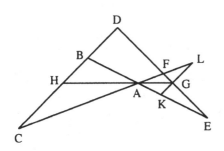 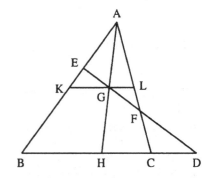

Let KL be drawn through G parallel to BD, and let it meet AB, AC in K, L. Now the rectangle BH, CD is to the rectangle BD, CH as the rectangle EG, FD is to the rectangle ED, GF; moreover the ratio of the rectangle EG, FD to the rectangle ED, GF is compounded from the ratio which GE has to ED, that is the ratio KG to BD, and the ratio which DF has to FG, that is the ratio CD to GL; and the ratio of the rectangle BH, CD to the rectangle BD, CH is compounded from the ratio HB to BD and the ratio DC to CH; thus the ratio compounded from the ratio KG to BD and the ratio CD to GL

will be the same as that compounded from the ratio HB to BD and the ratio
DC to CH. But the ratio KG to BD is compounded from the ratio KG to
HB and the ratio HB to BD; therefore the ratio compounded from the ratio
KG to HB and the ratio HB to BD along with the ratio CD to GL is the
same as that compounded from the ratio HB to BD and the ratio DC to CH.
Let the common ratio HB to BD be removed; therefore the remaining ratio
which is compounded from the ratio KG to HB and the ratio CD to GL is
the same as the ratio DC to CH, that is the ratio compounded from the ratio
DC to GL and the ratio GL to CH. Again let the common ratio DC to GL
be removed; therefore the remaining ratio KG to HB is the same as the ratio
GL to CH; therefore, *permutando*, BH is to HC as KG is to GL. But KL,
BH are parallel to each other; therefore there is a straight line which passes
through the points H, A, G. *

188)

Proposition 28

(This is Proposition 137 of Pappus's Book 7,
and his eleventh Lemma for the Porisms.)

We shall show the things which relate to its cases † similarly to the things
which have already been said, of which it is the converse, namely:
"Let ABC be a triangle, and let AD be drawn parallel to BC, and DE having
been drawn let it meet BC in the point E 'and AB, AC in F, G'. I say that
CB is to BE as the rectangle ED, FG is to the rectangle EF, GD."

Let CH be drawn through C parallel to DE, and let AB be produced to
H; and so since CH is to GF as CA is to AG, and ED is to DG as CA is to
AG, then CH will be to FG as ED is to DG, and consequently the rectangle
CH, GD is equal to the rectangle ED, FG. But the other specified rectangle is

---

* This will be clear from the following Lemma, viz.

Let ABC be a triangle, and let DE be drawn parallel
to BC, and let DE, BC be cut in the same ratio at F, G;
the points A, F, G will be in a straight line.

Since BG is to GC as DF is to FE, *componendo*, BC
will be to CG as DE is to EF; but AC is to CB as AE
is to ED; therefore, *ex aequali*, AC is to CG as AE is to
EF. And EF, CG are parallel; therefore the points A, F,
G are in a straight line.

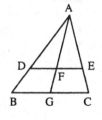

† Proposition 137 of Pappus's Book 7 is the case of Proposition 129, that is
Proposition 17 of the present work, in which to three straight lines meeting
in the same point, viz. AB, CA, AD, two straight lines ED, EC are drawn
of which one EC is parallel to one of the three AD; hence the rectangle
contained by BC and the segment between E and the intersection of AD, EC
is to the rectangle contained by EB and the segment between C and the same
intersection as the straight line CB is to the straight line BE.

EF, GD; therefore as the rectangle ED, FG is to the rectangle EF, GD so the rectangle CH, GD is to the rectangle EF, GD, that is, so CH is to EF, and so CB is to BE. It follows from these things that CB is to BE as the rectangle ED, FG is to the rectangle EF, GD.

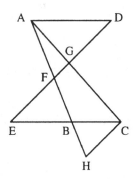

In Commandino's edition is found: "In the same way, and if the parallel AD is drawn on the other side, and" &c. up to the end of the Proposition, according as he considers the Greek text should be restored and supplemented; these things are entirely superfluous, as also is Commandino's demonstration at note G in his commentary; for if in his second figure C is put in place of B, and B in place of C, it will be the same case, and the same demonstration as the preceding one; as also if the point E is between B and C.

(p.188) ## Proposition 29

(This is Proposition 138 of Pappus's Book 7,
and his twelfth Lemma for the Porisms.)

These things having been demonstrated, the following now has to be shown: "Suppose that AB, CD are parallel, and that certain straight lines AD, AF, BC, BF are drawn between them, 'of which AD, BC meet in M', and let EC, ED be joined; there is a straight line which passes through the points G, M, K."

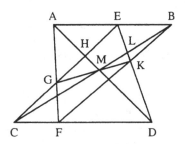

This is enunciated in more detail as follows:
"Suppose that between two parallel straight lines AB, CD two straight lines AD, BC are drawn, and from the points A, B which are on one of the parallel

straight lines let two straight lines AF, BF be inflected to the other CD; and from the remaining points C, D let CE, DE be inflected to AB; the straight line GK which passes through the points of intersection G, K of the inflected lines which are made by the straight lines which are not inflected from the extremities of the same drawn line, will pass through the point M, namely the point of intersection of the drawn lines AD, BC."

For since DAF is a triangle, and AE is parallel to DF, and EC is drawn meeting DF in C and AF, AD in G, H, by the preceding Lemma, namely Proposition 28, the rectangle CE, GH will be to the rectangle CG, HE as DF is to FC. Again since CBF is a triangle, and BE is parallel to CD, and ED is drawn meeting CF in D and BC, BF in L, K, the rectangle DE, KL will be to the rectangle DK, LE as CF is to FD, and *invertendo*, the rectangle DK, LE will be to the rectangle DE, KL as DF is to FC. But the rectangle CE, GH was to the rectangle CG, HE as DF is to FC; therefore the rectangle DK, LE is to the rectangle DE, KL as the rectangle CE, GH is to the rectangle CG, HE. And so the matter is reduced to Pappus's tenth Lemma, or Proposition 26 of this work: since the two straight lines EC, ED have been drawn to the two straight lines CML, DMH, and the rectangle DK, LE is to the rectangle DE, LK as the rectangle CE, GH is to the rectangle CG, HE, there will be a straight line which passes through G, M, K. For this has been demonstrated.

.188)

Proposition 30

(This is Proposition 139 of Pappus's Book 7, and his thirteenth Lemma.)

"Now suppose that AB, CD are not parallel, and let them meet in the point N, the rest remaining as in the preceding Proposition; again I say that there is a straight line which passes through the points G, M, K."

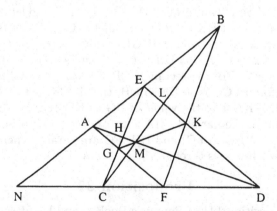

For since the two straight lines CE, CD have been drawn from the same point C to the three straight lines AN, AF, AD, the rectangle NC, FD is to the rectangle ND, CF as the rectangle CE, GH is to the rectangle CG, HE [Prop. 17]. Again since from the same point D the two straight lines DE, DN have

been drawn to the three straight lines BN, BC, BF, the rectangle DK, LE
will be to the rectangle DE, KL as the rectangle NC, FD is to the rectangle
ND, CF [Prop. 17]. But it has been shown that the rectangle CE, GH is to
the rectangle CG, HE as the rectangle NC, FD is to the rectangle ND, CF.
Therefore the rectangle DK, LE is to the rectangle DE, KL as the rectangle
CE, GH is to the rectangle CG, HE. And the matter is reduced to the same
Lemma as in the case of parallel lines. Therefore from what has already been
said there is a straight line which passes through G, M, K.

Proposition 31

(p.188)

(This is Proposition 140 of Pappus's Book 7, and his fourteenth Lemma for
the Porisms; it is also the converse of Proposition 28 of the present work.)

"Let AB be parallel to the straight line CD, let AE, BC be drawn, and let
F be the point on the straight line BC such that the rectangle CB, GF has
the same ratio to the rectangle CG, FB as DE has to EC. I say that there is
a straight line which passes through the points A, F, D."

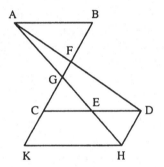

Indeed let DH be drawn through D parallel to BC, and let AE be produced
to H; now through H let HK be drawn parallel to CD, and let BC be produced
to K. And so since the rectangle CB, GF is to the rectangle CG, FB as DE is
to EC, and as DE is to EC so DH is to CG and so the rectangle DH, FB is
to the rectangle CG, FB, the rectangle CB, GF will be equal to the rectangle
DH, FB; therefore as CB is to BF so DH, that is CK, is to GF, and because
of it the whole of KB is to the whole of BG as KC, that is DH, is to GF (n).
But, on account of the parallels, HA is to AG as KB is to BG; consequently
HA is to AG as DH is to FG; and DH, FG are parallel; therefore there is a
straight line which passes through the points A, F, D.

Proposition 32

(This is Proposition 141 of Pappus's Book 7, and his fifteenth Lemma.)

"This having been set forth, let AB be parallel to the straight line CD, and
let the straight lines AF, FB and CE, ED be drawn between them; and let

BC, GK be joined 'which meet in M'. I say that there is a straight line which passes through the points A, M, D."

This is enunciated in more detail as follows:
"Let AB be parallel to the straight line CD, and let the straight lines AF, BF be inflected from the points A, B to CD; then let CE, DE be inflected from the points C, D to AB, and let G be the point of intersection of AF, CE, and let K be the point of intersection of the remaining pair BF, DE; and let BC be drawn which meets the join GK in M; the points A, M, D will be in a straight line."

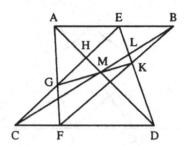

Let DM be joined and let it be produced to H; therefore since BE has been drawn from the vertex B of the triangle BCF parallel to CD, and ED has been drawn from the point E on BE, meeting BC in L, the rectangle DE, KL is to the rectangle DK, LE as CF is to FD [Prop. 28]. Now the rectangle CG, HE is to the rectangle CE, GH as the rectangle DE, KL is to the rectangle DK, LE [Prop. 17], for the two straight lines EC, ED have been drawn from the same point E to the three straight lines CL, DH, GK which meet in the same point M. Therefore, *invertendo*, the rectangle CE, GH is to the rectangle CG, HE as DF is to FC. Consequently, by the preceding Proposition, there is a straight line which passes through the points A, H, D; but the point M is on the straight line DH, therefore the points A, M, D are also in a straight line.

Proposition 142 of Pappus's Book 7, which is his sixteenth Lemma, is exactly the same as his Proposition 136, but it has another demonstration, as has been said, which has already been given in Proposition 27 of this work.

Proposition 33

(This is Proposition 143 of Pappus's Book 7, and his seventeenth Lemma.)

"Now suppose that AB is not parallel to CD, but let them meet at the point N, 'the rest remaining as in the preceding Proposition'."

'Let DM be joined and let it be produced to H'; therefore since the two straight lines 'DN, DE' have been drawn from the same point D to the three straight lines BN, BC, BF, the rectangle DE, KL is to the rectangle DK, LE as the rectangle CF, DN is to the rectangle CN, FD [Prop. 17]. But the rectangle CG, HE is to the rectangle CE, GH as the rectangle DE, KL is to the rectangle

DK, LE [Prop. 17]; for again the two straight lines EC, ED have been drawn from the same point E to the three straight lines CL, DH, GK which meet in the same point M. Therefore the rectangle CF, DN is to the rectangle CN, DF as the rectangle CG, HE is to the rectangle CE, GH. Therefore by Proposition 26 there is a straight line which passes through A, H, D,* and consequently there is a straight line which passes through the points A, M, D.

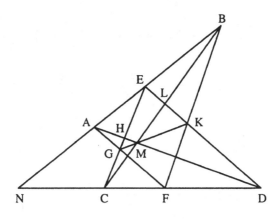

(p.188)

Proposition 34

(This is a Porism, namely one of those among the Porisms of
Euclid's Book 1, which Pappus presented in these words,
"That this goes to a given point".)

"Suppose that three straight lines which meet in the same point are given in position, and from two points which are given on one of them two straight lines are inflected to another straight line which is given in position; then the straight line which joins two points of intersection of these and the remaining two of the three straight lines which are given in position passes through a given point."

Let there be three straight lines AF, BF, CFD which are given in position and meet in the same point F, and let two points C, D be given on one of them, from which two straight lines CE, DE are inflected to a fourth straight line AB which is given in position, and let G, K be two of the points of intersection of CE, DE and the straight lines AF, BF; the join GK will pass through a given point.

First (Fig. 1), let the fourth straight line AB be parallel to the straight line CD on which the two points are given, and let A, B be the points of intersection of AB and FA, FB. To one of these, say B, let straight lines

---

* For the two straight lines CN and CGE have been drawn from the same point C to the two straight lines AN, AF, and two points D, H have been taken on CN, CGE which make the last stated proportion.

CB, DB be inflected from the points C, D, and since the point B is the point of intersection of FB, DB, while the point of intersection of FA, CB is on CB, the straight line CB will join two points of intersection of the inflected straight lines CB, DB and FA, FB; therefore, by hypothesis, CB passes through the given point. Again from the points C, D let straight lines CE, DE be inflected to AB in any manner, and let G, K be the points of intersection of them and the straight lines FA, FB; by hypothesis, therefore, the join GK passes through the given point. Let GK meet CB at M, therefore the point M is that point which the Porism asserts to be given, which is in fact clear from Proposition 32, for it has been shown in it that the point M is on the straight line AD, but the same point is also on the straight line CB, and AD, CB are given in position, for the points A, B, C, D are given; therefore M is the given point through which the straight line GK passes.

Therefore let AD, CB be joined and let them meet at M; then M will be the point which has to be found; that is to say, if straight lines CE, DE are inflected in any manner to AB from the points C, D, and they meet FA, FB in G, K, then the points G, M, K will be in a straight line, as follows from Proposition 29.

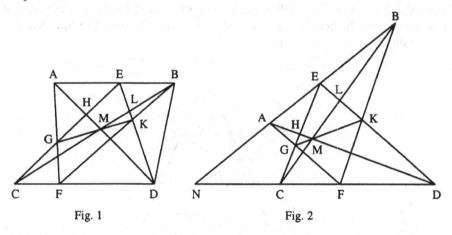

Fig. 1   Fig. 2

But suppose (Fig. 2) that AB is not parallel to CD, the rest remaining as in the first case, and let M be the point of intersection of CB, GK; therefore A, M, D will be in a straight line by Proposition 33, and since the straight lines CB, AD are given in position, the point M will be given. Therefore if straight lines CE, DE are inflected in any manner to the straight line AB from the points C, D, and they meet FA, FB in G, K, the join GK will pass through the point M; for the points G, M, K are in a straight line by Proposition 30.

But this Porism can be shown by use of many fewer Lemmas as follows.

Proposition 35

"Suppose that on two parallel straight lines AB, CD two points A, C are taken, A being on one and C being on the other, and that two straight lines AE, CE are inflected from them to any straight line BD which meets the parallel straight lines; then the rectangle GB, FD contained by the segments of the parallel lines between the inflected lines and the straight line BD will be equal to the rectangle AB, CD contained by the segments of the same straight lines between the points A, C and the same straight line."

For, on account of the parallels, FD is to DC as AB is to BG. Therefore the rectangle GB, FD is equal to the rectangle AB, CD. *

(p.189) ## Proposition 36

"In the same situation as before suppose in addition that another two straight lines BH, DH are inflected from the points B, D in which the straight line BD meets the parallel lines to the straight line which is drawn through the points A, C; and let L be the point of intersection of the one BH which is inflected from the point B on AB and the straight line CE which is inflected from the point C on the other parallel line CD, and let K be the point of intersection of the remaining inflected lines DH, AE; then the join KL will be parallel to the parallel lines."

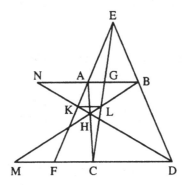

Let the inflected lines BH, DH meet the parallel lines in the points M, N; and since the straight lines BH, DH have been inflected from the points B, D on the parallel lines AB, CD to the straight line AC, meeting the parallel lines in M, N, by Proposition 35 the rectangle AN, CM will be equal to the rectangle AB, CD, to which the rectangle GB, FD has been shown to be equal. Therefore MC is to GB as FD is to NA and so, on account of the parallels, ML is to LB as DK is to KN; and *componendo*, MB is to BL as DN is to NK; but HB is to BM as HN is to ND; therefore, *ex aequali*, HB is to BL as HN is to NK. Therefore the straight line KL is parallel to the straight line NB.

---

* See the figure for Proposition 36.

Proposition 37

"Suppose that two straight lines AC, BD are drawn between two straight lines AB, CD, and from two of their extremities A, B, which are on one of the straight lines AB, CD, let two straight lines AE, BE be inflected to the other straight line CD; and let two other straight lines CF, DF be inflected from the remaining extremities C, D of the drawn lines to the straight line AB; the straight line which passes through the two points of intersection of the inflected lines which are made by the straight lines which are not inflected from the extremities of the same drawn line, will pass through the point of intersection of the drawn lines."

Let G be the point of intersection of the straight lines AE, DF, and H the point of intersection of the remaining inflected lines BE, CF, and let K be the point of intersection of the drawn lines AC, BD; there will be a straight line which passes through the points G, K, H.

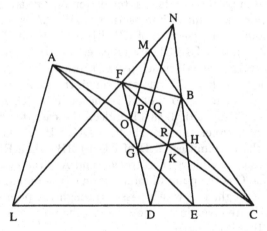

Let BC be joined, and through the point A, namely an extremity of one of the drawn lines AC, let AL be drawn parallel to the remaining drawn line BD, and let it meet CD in L; now let the join LF meet CB in M and BE in N; next let O be the point of intersection of the straight lines AC, DF, and let the joins OM, GN meet CF in the points P, Q; finally let BD meet the same CF in R.

Therefore since AL, BD are parallel, and AE, BE have been inflected from their extremities A, B to the straight line CD, and LF, DF have been inflected from the remaining extremities L, D to the straight line AB, and G, N are their intersections, which are in fact made by straight lines which are not inflected from points in the same parallel, then, by the preceding Proposition, the straight line GN will be parallel to the straight lines AL, BD; and similarly, since the straight lines AC, BC have been inflected to CD from the extremities A, B of the parallels, meeting the inflected lines DF, LF in O, M, the straight line OM will be parallel to the same AL, BD, and consequently to GN. Therefore, on account of the parallels, KR is to RB as (OP is to PM, that is as) GQ is to QN; and *permutando*, as KR is to GQ so (RB is to QN, and so) RH is to HQ. And KR, GQ are parallel; therefore there is a straight line which passes through the points H, K, G.

The preceding 34th Proposition is established
by means of this Proposition as follows.

Let three straight lines AE, BE, DEC which meet in the same point E be given in position, and on one of them let the two points C, D be given, from which two straight lines CF, DF are inflected to a fourth straight line AB which is given in position, and let G, H be two of the points of intersection of CF, DF and the straight lines AE, BE; the join GH will pass through a given point.

For the straight lines CB, DB have been inflected to the point B in which the straight lines AB, EB meet, therefore, by hypothesis, the straight line DB which passes through the points of intersection of the straight lines DB, CB and EA, EB, (for the point of intersection of DB, EA is on the straight line DB, and the point of intersection of CB, EB is the point B,) passes through the given point; again if DF, CF are inflected to AB in any manner, by hypothesis, the straight line GH which passes through the points of intersection of them and the same EA, EB, passes through the same given point; let DB, GH meet each other in K, so that K is the point which has been asserted to be given in the Proposition; this is indeed the case, for since GKH, DKB have been drawn between the straight lines DG, HB, and GE, DE have been inflected from the extremities G, D of DG to HB, while HF, BF have been inflected from the extremities H, B to DG, and A is the point of intersection of GE, BF, while C is the point of intersection of the remaining pair DE, HF, there will be a straight line which passes through the points A, K, C [Prop. 37]. But AC, DB are given in position, and so the point K through which the straight line GH passes is given.

Therefore let AC, BD be joined and let them meet in K; then K will be the point which has to be found; that is to say, if straight lines CF, DF are inflected from the points C, D to AB in any manner, and they meet the straight lines EA, EB in G, H, then the points G, K, H will be in a straight line, as is established from Proposition 37.

The converse of the 34th Proposition is the following.

(p.189) **Proposition 38**

"Suppose that three straight lines AE, BE, DEC which meet in the same point E are given in position, and let two points C, D be given on one of them, and let a point K be given outside of these lines through which a straight line GK is drawn in any manner to meet EA, EB in G, H; if CH, DG are joined and meet in F, the point F will lie on a straight line which is given in position."

In order that the enunciation be more like the enunciation of the main part of Pappus's Proposition on the four lines, it can be expressed as follows, viz.

"Suppose that four straight lines DC, GH, DF, CF meet each other and that not more than two pass through the same point, and on one of them let two points of intersection D, C be given, and let another GH pass through a given point K which is outside of the straight line DC; suppose further that the intersections G, H of GH with the remaining straight lines DF, CF lie on straight lines EA, EB which are given in position and whose point of intersection E is on the straight line DC; then the point F will lie on a straight line which is given in position."

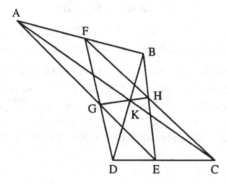

Let DK be joined and let it meet EB in B; and since the points D, K are given, the straight line DB is given in position; and EB is given in position, so that the point B is given. Let the straight line GH be drawn through the point K in any manner, meeting EA, EB in G, H, and let the joins DG, CH meet in F, and let the join BF meet EA in A. Now GKH, DKB have been drawn between the straight lines GD, HB, and GE, DE have been inflected from the points G, D to HB, while HF, BF have been inflected from the points H, B to the straight line GD, and the points A, C are the points of intersection of the inflected lines which are not made by straight lines inflected from the extremities of the same drawn lines, and the point K is the point of intersection of the drawn lines GH, DB. Thus, by Proposition 29 or 30, the points A, K, C will be in a straight line. But the points K, C are given, so that the straight line AC is given in position, and EA is given in position; therefore the point A is given, and B is given; therefore the straight line AB on which the point F lies is given in position.

Therefore let DK be joined and let it meet EB in B; and let the join CK meet EA in A; the join AB will be the straight line on which the point F lies; that is to say, if any straight line GKH is drawn through the point K, meeting the straight lines EA, EB in G, H, and DG, CH are joined and meet in F, then the point F will be on the straight line AB. For since DG, CH have been drawn between the two straight lines DC, GH, and the straight lines DK, CK have been inflected from the points D, C, which are on one of the straight lines, to the remaining one GH, and the straight lines GE, HE have been inflected from the points G, H on the remaining straight line to the first straight line DC, and A, B are the points of intersection of the inflected lines which are not made by straight lines which are inflected from the extremities of the same drawn line, the join AB will pass through F, the point of intersection of the drawn lines DG, CH, that is the point F is on the straight line AB.

And, with the same construction, if straight lines CF, DF are inflected from the points C, D to AB, meeting EA, EB in G, H, there will be a straight line which passes through the points G, K, H; this is in fact clear from Proposition 37.

(p.189)

Proposition 39

(This is Proposition 144 of Pappus's Book 7,
whose enunciation has been made more clear.)

"Let ABC be a triangle, and let AD be parallel to BC, and let DE be drawn in any manner to meet AB, AC in H, F. Moreover let G be the point on BC which makes the square of EB to the rectangle EC, CB as the straight line BG is to the straight line GC, and let FG be joined, which the join BD meets in K; I say that there is a straight line which passes through the points H, K, C."

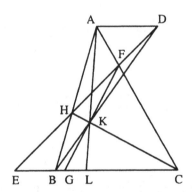

For since BG is to GC as the square of EB is to the rectangle EC, CB, let the ratio CE to EB, which is in fact the same as the ratio of the rectangle EC, CB to the rectangle EB, BC, be introduced on both sides, and "the ratio compounded of the ratio of the square of EB to the rectangle EC, CB and the ratio of the rectangle EC, CB to the rectangle EB, BC, that is" the ratio of the square of EB to the rectangle EB, BC, that is the ratio of the straight line EB to the straight line BC, will be the same as that which is compounded of the ratios BG to GC and CE to EB, which is the same as the ratio of the rectangle BG, CE to the rectangle GC, EB. Moreover, by Proposition 28 of this work, the rectangle DF, HE is to the rectangle DE, FH as EB is to BC; therefore the rectangle DF, HE is to the rectangle DE, FH as the rectangle BG, CE is to the rectangle GC, EB. Therefore there is a straight line which passes through the points H, K, C; for this has been demonstrated in Proposition 27 of this work in those things which relate to the converse case. For two straight lines EBG, EFD have been drawn from the same point E to the two straight lines KB, KF and points C, H have been taken on them which make the rectangle BG, CE to the rectangle GC, EB as the rectangle DF, HE is to the rectangle DE, FH.

N.B. The converse of Proposition 39 is the following.

"Let ABC be a triangle, and let AD be parallel to BC; if any straight line DFHE is drawn and BD, CH are joined, meeting each other in K, and the join FK meets BC in G, then the square of EB will be to the rectangle EC, CB as the straight line BG is to the straight line GC."

For since AD is parallel to the straight line BC, and DFHE has been drawn, by Proposition 28 the rectangle EH, FD will be to the rectangle ED, HF as EB is to BC. And since the two straight lines ED, EC have been drawn from the same point to the three straight lines KB, KH, KF, which are drawn from the same point K, the rectangle EC, BG will be to the rectangle EB, GC as (the rectangle EH, FD is to the rectangle ED, HF by Proposition 17, that is as) the straight line EB is to the straight line BC. Therefore since the straight line EB is to the straight line BC as the rectangle EC, BG is to the rectangle EB, GC, let the ratio EB to EC be introduced on both sides, and the ratio of the square of EB to the rectangle EC, CB will be the same as (the ratio compounded from the ratio of the rectangle EC, BG to the rectangle EB, GC and the ratio of the straight line EB to the straight line EC, that is the ratio of the rectangle EB, GC to the rectangle EC, CG, that is) the ratio of the rectangle EC, BG to the rectangle EC, CG, that is the ratio of the straight line BG to the straight line GC.

This demonstration is a little more detailed than that given by Pappus, namely by the addition of what is contained between the inverted commas; also Pappus's text in Commandino's edition twice has erroneously the rectangle DE, FH in place of the rectangle DF, HE, and conversely; these things have now been corrected.

.189)

Proposition 40

(This is a Porism to which the preceding Proposition applies.)

"Suppose that there are five straight lines AB, AC, AD, DE, EC, the first three of which meet in the single point A, and that two of them AD, EC are parallel and on that one EC which does not pass through the point A three points of intersection E, B, C are given; moreover, of the remaining points of intersection A, D, F, H let two D, H lie on straight lines BK, CK which are given in position and which pass through two of the given points B, C; then the remaining points of intersection A, F will lie on straight lines which are given in position."

For there are four straight lines EB, BA, AD, DE of which two EB, AD are parallel to each other, and two points E, B are given on one of the parallel lines, while of the remaining points of intersection A, D, H two D, H lie on straight lines BK, CK which are given in position, and one of them BK passes through one of the given points B; thus by Case 3 of Proposition 8 of this work (n1) the remaining point of intersection A will lie on a straight line which is given in position. Let it be AK which meets the straight line EC in L. Therefore, by what has been shown in that case, EC will be to CB as BC is to CL. Again since there are four straight lines EC, ED, DA, AC of which two EC, AD are parallel to each other, and two points E, C are given on one of the parallel lines, while of the remaining points of intersection A, D, F two A, D lie on straight lines LK, BK which are given in position, by Article 2 in Case 4 of Proposition 8 of this work (n2) the remaining point of intersection F will lie on a straight line which is given in position. Let it be FK, and let it meet the straight line EC in G. And, from what has been shown in that case, EC will be to BL as CG is to GL. Therefore the points A, F lie on straight lines which are given in position.

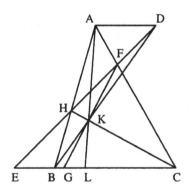

It will be composed thus.

Let BC be made to CL as EC is to CB, and let LK be joined; this will be the straight line on which the point A lies, as has been shown in Case 3 of Proposition 8. Then let CG be made to GL as EC is to BL, and the join GK will be the straight line on which the point F lies, as has been shown in Article 2 of Case 4 of Proposition 8 of this work.

Now it seems that Euclid combined these two constructions into one for finding the point G, doubtless by making the straight line BG to the straight line GC as the square of EB to the rectangle EC, CB. That this construction may be derived from the two described will be shown as follows (n3).

Since EC is to CB as BC is to CL, *dividendo*, EB will be to BC as BL is to LC; and *permutando*, and by [VI 1] and *permutando* again, the rectangle

EB, BL is to the rectangle EC, CL as the square of EB is to the rectangle BC, CE. And since, in the second construction, EC was to BL as CG is to GL, then EG will be to GB as CG is to GL [V 19], and, *dividendo*, EB will be to BG as CL is to LG. Therefore the rectangle EB, GL is equal to the rectangle BG, LC. And since EC is to BL as CG is to GL, *invertendo*, BL is to EC as LG is to GC; thus *permutando*, and by [VI 1] and *permutando* again, as the rectangle EB, BL is to the rectangle EC, CL so (the rectangle EB, LG is to the rectangle LC, CG, that is so the rectangle BG, LC is to the rectangle LC, CG, and so) the straight line BG is to the straight line GC. But the square of EB was to the rectangle BC, CE as the rectangle EB, BL is to the rectangle EC, CL; therefore the straight line BG is to the straight line GC as the square of EB is to the rectangle BC, CE.

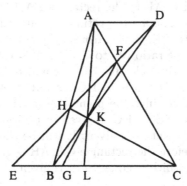

It will be composed as follows.

Therefore when the point G has been found by this construction, the join GK will be the straight line on which the point F lies, that is to say, if straight lines EF, CF are inflected from the points E, C to the straight line GK in any manner, if from the point D in which the straight line EF meets the straight line BK the straight line DA is drawn parallel to the straight line EC to meet the straight line CF in A, and if the join AB meets the join ED in H, then the points H, K, C will be in a straight line by Proposition 39.

Proposition 41

(This is a Porism, namely one of those which Pappus gives among the Porisms of Euclid's Book 1 with these words "That the straight line ... takes off from those given in position segments containing a given rectangle.")

"Suppose that two straight lines have been given in position; if a straight line is drawn in any manner from a given point off them, it will take off from the straight lines which are given in position segments which are adjacent to given points on them and which contain a given rectangle."

Let the straight lines AB, AC be given in position, and from the given point D let a straight line BDC be drawn in any manner, meeting them in

B, C; and suppose that the Porism is true, namely the segments BG, CH cut off by the straight line BC and adjacent to given points G, H contain a given rectangle BG, CH. Let any other straight line EDF be drawn through the point D, and let AD be joined; therefore, by hypothesis, both the rectangle EG, FH and the rectangle AG, AH are equal to the same given rectangle, and consequently to the rectangle BG, CH. Therefore (n) BG is to FH as EG is to CH, and hence BE is to FC as BG is to FH. Also AH is to EG as FH is to AG, and hence FH is to AG as AF is to AE. Now, since the straight lines AB, BC meet the straight lines AF, FE, by the Lemma at the end of Proposition 22 the ratio BD to DC is the same as the ratio compounded from the ratio BE to EA and the ratio AF to FC, that is the ratio compounded from the ratio BE to FC and the ratio AF to EA, for both ratios are the same as the ratio of the rectangle BE, AF to the rectangle EA, FC [VI 23]; therefore the ratio BD to DC is the same as (the ratio compounded from the ratio BE to FC and the ratio AF to AE, that is from the ratio BG to FH and the ratio FH to AG, that is as) the ratio BG to GA. Therefore the join GD is parallel to AF. And since the rectangle BG, HC is equal to the rectangle AG, AH, then BG is to GA, that is BD is to DC, as AH is to HC; therefore DH is parallel to BA. Therefore since the point D is given, and DG is parallel to AF which is given in position, DG is given in position, and AB is given in position, therefore the point G is given. It will be shown similarly that the point H is given. Therefore the rectangle GA, AH is given, and consequently the rectangle BG, CH, which is equal to it, is given.

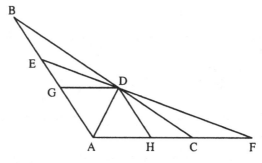

It will be composed thus.

Through the point D let DG, DH be drawn parallel to AC, AB; then G, H will be the points which had to be found and GA, AH will be the rectangle which had to be found. For let any straight line BDC be drawn through D; and since DH is to HC as BG is to GD, the rectangle BG, HC is equal to the rectangle GD, DH, that is to the rectangle HA, AG.

Proposition 42

(This is Proposition 148 of Pappus's Book 7, and his Lemma 22.)

"Let AB be a straight line, and let two points C, D be taken on it; and let twice the rectangle contained by AB, CD be equal to the square of CB. I say that the square of AD is equal to the squares of AC, DB."

```
A       C   D       B
———————————————————————
```

For since twice the rectangle contained by AB, CD is equal to the square of CB, let the rectangle contained by twice BD, DC be taken away from both sides, and the remaining part, twice the rectangle contained by AD, AC, will be equal to the squares of CD, DB. Again let the square of CD be taken away from both sides, therefore the remaining part, namely twice the rectangle contained by AC, CD along with the square of CD, is equal to the other remaining part which is the square of DB. Let the square of AC be added to both sides; therefore the whole part, which is the square of AD, will be equal to the squares of AC, DB.

And the converse of the Proposition will be shown by working backwards.

90) # Proposition 43

(This is Proposition 149 of Pappus's Book 7 and his Lemma 23.)

"Let the rectangle AB, BC be equal to the square of BD. I say that three things hold, namely: the rectangle which is contained by both AD, DC, and BD is equal to the rectangle AD, DC; then the rectangle contained by both AD, DC, and CB is equal to the square of DC; finally the rectangle contained by both AD, DC, and AB is equal to the square of AD."

```
A       C   B       D
———————————————————————
```

For (n) since the rectangle AB, BC is equal to the square of BD, then, on account of proportion, * and the whole part to the whole part, † and *invertendo*, and *componendo*, CD will be to DB as both CD, DA to DA. Therefore the rectangle which is contained by both AD, DC, and BD is equal to the rectangle AD, DC. Again since the whole of AD is to the whole of DC as DB is to BC, *componendo*, DC will be to CB as both AD, DC to DC. Hence the rectangle contained by both AD, DC, and CB is equal to the square of DC. Again since the whole of AD is to the whole of DC as AB is to BD, *invertendo* and *componendo*, DA will be to AB as both CD, DA to DA. Therefore the rectangle which is contained by both AD, DC, and AB is equal to the square of AD.

---

* AB is to BD as DB is to BC.

† AD is to DC as DB is to BC.

Proposition 44

(This is Proposition 150 of Pappus's Book 7 and his Lemma 24.)

"Let AB be a straight line, and C, D two points on it; and let the square of CD be equal to twice the rectangle contained by AC, DB. I say that the square of AB is equal to the squares of AD, CB."

For since the square of CD is equal to twice the rectangle which is contained by AC, DB, twice the rectangle contained by AC, CB will be equal to the square of CD and twice the rectangle contained by AC, CD. Let the square of AC be added to both sides, therefore twice the rectangle contained by AC, CB along with the square of AC is equal to the square of AD. Again let the square of BC be added to both sides; therefore the whole square of AB will be equal to the squares of AD, CB.

And the converse will be shown by working backwards.

(p.190)

Proposition 45

(This is Proposition 151 of Pappus's Book 7, and his Lemma 25.)

"Let the rectangle AB, BC be equal to the square of BD. I say that three things hold, namely: the rectangle which is contained by the excess of AD, DC, and BD is equal to the rectangle AD, DC; then the rectangle contained by the excess of AD, DC, and CB is equal to the square of DC; finally the rectangle contained by the excess of AD, DC, and BA is equal to the square of AD."

Fig. 1   Fig. 2

For since DB is to BC as AB is to BD, the remaining part will be to the remaining part, (n) "viz. AD will be to DC as AB is to BD", and *dividendo* in Fig. 1 or *dividendo inverse* in Fig. 2, AD will be to DB as the excess of AD, DC is to DC. Therefore the rectangle which is contained by the excess of AD, DC, and DB is equal to the rectangle AD, DC. Again since the remaining part AD is to the remaining part DC as DB is to BC, *dividendo* or *dividendo inverse*, DC will be to CB as the excess of AD, DC is to DC. Therefore the rectangle contained by the excess of AD, DC, and CB is equal to the square of DC. Again since AB is to BD as AD is to DC, *convertendo*, AB will be to AD as AD is to the excess of AD, DC. Hence the rectangle which is contained by the excess of AD, DC, and AB is equal to the square of AD. Or if BC is greater than BA, as in Fig. 2, since AD is to DC as AB is to BD, the same will follow *invertendo* and *dividendo*.

This third part of the demonstration is now presented unimpaired; for it seems to be corrupted in the Greek codices, as it is in Commandino's edition, and it has been ineptly explained in his Commentary at note D.

Proposition 46

(This is not included in Commandino's edition,
but it is necessary for the subsequent Porism.)

"Let AB be a straight line, and let C, D be two points on it, of which C is on AB produced, while D is between the points A, B; and let twice the rectangle contained by AB, DC be equal to the square of AC, that is to the squares of AB, BC and twice the rectangle AB, BC; then the square of DB will be equal to the squares of AD, BC."

A   D   B   C

For twice the rectangle contained by AB, DC, that is twice the rectangle contained by AB, BD along with twice that contained by AB, BC, is equal to the square of AC, that is to the squares of AB, BC and twice the rectangle AB, BC; let twice the rectangle contained by AB, BC be taken away from both sides, and the remaining part, twice the rectangle contained by AB, BD, will be equal to the squares of AB, BC. Again let twice the rectangle AD, DB along with the square of DB be taken away from both sides; the remaining part, the square of DB, will be equal to the remaining part, namely the squares of AD, BC.

And the converse will be shown by working backwards.

190)
Proposition 47

(This is a Porism for which Proposition 148 and
Proposition 150 of Pappus's Book 7 are useful.)

"Suppose that a point C is taken on a straight line AB, either between the points A, B, or on AB produced on either side; a fourth point D will be given on the same straight line, such that if any point E is taken on AB produced on the side of B, the square of the segment between it and the point D which has to be found will be equal to the square of the segment between the same point E and the point C, along with the square of the segment between the point D which has to be found and the point A, and the rectangle contained by the segment between the point E and the point B and a certain given straight line."

Case 1. Here the point C is between the points A, B.

G A   D H   C   B   E

Suppose that the Porism is true, and let D be the point and F the straight line which have to be found. Therefore if a point E is taken in any manner on

AB produced, by hypothesis, the square of DE will be equal to the squares of CE, AD along with the rectangle F, BE. Again let the point B be taken in place of the point E, and, by hypothesis, the square of DB will be equal to the squares of CB, AD. Therefore (n1) the excess of the square of DE over the square of DB will be equal to the excess of the squares of CE, CB along with the rectangle F, BE; that is, if GD is made equal to DB, and HC to CB, the rectangle GE, EB will be equal to the rectangle HE, EB along with the rectangle F, BE [II 6]. Therefore the straight line GE is equal to the straight lines HE, F together; and if HE is taken away from both, the remaining part GH will be equal to F. Now since GB is twice DB, and HB is twice CB, then GH will be twice DC, so that the straight line F is also twice the same DC. And since the square of DB is equal to the squares of CB, AD, twice the rectangle AB, DC will be equal to the square of AC by the converse of Proposition 42 (n2). But the straight lines AB, AC are given, therefore DC is given; and the point C is given, so that the point D is also given; now the straight line F is twice the straight line DC; therefore the straight line F is given. These are the things which will have to be found.

It will be composed as follows.

G A   D H   C   B   E

As twice AB is to AC so let AC be made to CD which is located from the point C towards the side on which A is in relation to the point B; now let the straight line F be taken equal to twice the straight line DC. If the point E is taken in any manner on AB produced on the side of B, the square of DE will be equal to the squares of CE, AD along with the rectangle F, BE. For, by construction, twice the rectangle AB, DC is equal to the square of AC; therefore, by Proposition 42, the square of DB is equal to the squares of CB, AD. Let GD be made equal to DB, and HC to CB; consequently GH will be equal to twice DC, that is the straight line F, so that GE will be equal to HE, F together; and the rectangle GE, EB, that is twice the rectangle DB, BE along
† with the square of BE, will be equal to the rectangles HE, EB, F, EB, † that is twice the rectangle CB, BE, the square of BE, and the rectangle F, BE. But the square of DB is equal to the squares of CB, AD; when these equal quantities have been added in, the square of DE will be equal to the squares of CE, AD and the rectangle F, EB [II 4].

Corollary. However if the point E is taken not on AB produced on the side of B, but either between A and B, or on BA produced on the side of A, the rest remaining as in the Proposition, then the square of DE along with the rectangle F, BE, or along with twice the rectangle DC, BE, will be equal to the squares of CE, AD.

This can be investigated in the same way as the preceding Porism; but it can be easily shown from it as follows.

Case 1

A   D   C E B

Let the point E be between the points C, B. Therefore (n3) since it has been shown in the composition of the preceding Porism that the square of DB is equal to the squares of CB, AD, let the square of CB be taken away from both sides, and the square of DC along with twice the rectangle DC, CB will be equal to the square of AD. Let the square of CE be added to both sides, and the whole quantity, namely the square of DC along with twice the rectangle DC, CB and the square of CE, will be equal to the whole quantity, namely the squares of CE, AD; but the square of DC along with twice the rectangle DC, CE and the square of CE is equal to the square of DE [II 4]. Therefore the square of DE along with twice the rectangle DC, EB is equal to the squares of CE, AD.

Case 2

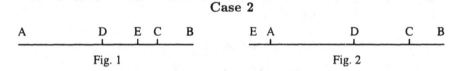

A   D   E C   B   E A   D   C   B

Fig. 1   Fig. 2

Let the point E be between the points A, C (Fig. 1), or on CA produced on the side of A (Fig. 2). Now (n4) it has been shown in Case 1 of the Corollary that the square of DC along with twice the rectangle DC, CB and the square of CE is equal to the squares of CE, AD; moreover the square of CD along with the square of CE is equal to the square of DE along with twice the rectangle DC, CE [II 7]; thus the square of DE along with twice the rectangles DC, CE and DC, CB, that is along with twice the rectangle DC, EB, will be equal to the squares of CE, AD.

Case 2 of Proposition 47. Here the point C is on BA produced on the side of A.

G H   D C A   B E

It will be shown exactly as in the analysis of Case 1 of the Proposition that the straight line F is twice the straight line DC, and the square of DB is equal to the squares of DA, CB; hence by the converse of Proposition 44 twice the rectangle DC, AB will be equal to the square of CA; for two points C, A have been taken on the straight line DB. But CA, AB are given, so that DC is given, and the point D will be given; therefore CD is given, and also the straight line F, which is twice CD.

It will be composed as follows.

As twice AB is to AC so let AC be made to CD which is located from the point C on AC produced on the side of C; now let the straight line F be taken equal to twice DC. If the point E is taken in any manner on AB produced on the side of B, the square of DE will be equal to the squares of CE, DA along with the rectangle F, BE. For, by construction, twice the rectangle DC, AB is equal to the square of CA; therefore, by Proposition 44, the square of DB is equal to the squares of DA, CB. And the rest of the composition will be expressed in the same words as that of Case 1 of the Proposition.

Corollary. However if the point E is taken either between the points C, B, or on the straight line BC produced on the side of C, the rest remaining as before, then the square of DE along with the rectangle F, BE, or twice the rectangle DC, BE, will be equal to the squares of DA, CE.

Case 1

$$D\;C\quad A\quad E\quad B$$

When the point E is between C, B, it is shown with the same words as Case 1 of the preceding Corollary.

Case 2

$$E\quad\;\;D\;C\quad A\qquad\;\;B$$

When the point E is on BC produced on the side of C, it is shown with the same words as in Case 2 of the preceding Corollary.

Case 3 of Proposition 47. Here the point C is on AB produced on the side of B.

$$G\;A\quad D\qquad B\qquad C\qquad H\;\;E$$

First (n5), if the straight line CE is bigger than the straight line CB, the analysis will be expressed in the same words as that of Case 1 of this Proposition; the same holds for the composition as far as the words "that is twice the rectangle CB, BE" etc. (see † above), in place of which let there be read: that is twice the rectangle CH, HE, the square of HE, and the rectangle F, BE; but the square of DB is equal to the squares of BC, AD, or of CH, AD; therefore the whole part, namely the square of DE, will be equal to the whole part, namely the squares of CE, AD and the rectangle F, BE. Moreover the same thing will be shown if CE is equal to CB.

$$G\;A\quad D\qquad B\qquad C\;E\quad H$$

Now (n6) suppose that the straight line CE is smaller than the straight line CB. By hypothesis, the square of DE is equal to the squares of CE, AD and the rectangle F, BE; and, if the point B is taken in place of the point E, by hypothesis, the square of DB is equal to the squares of BC, AD, that is if CH is made equal to CB, to the rectangle BE, EH and the squares of EC, AD [II 5]; therefore the excess of the squares of DE, DB will be equal to the excess of the rectangle F, BE over the rectangle BE, EH, that is, if GD is made equal to DB, the rectangle GE, EB will be equal to the excess of the rectangle F, BE over the rectangle BE, EH. Therefore the rectangle GH, BE is equal to the rectangle F, BE, so that the straight line GH is equal to the straight line F. And since the square of DB is equal to the squares of AD, BC, twice the rectangle AB, DC will be equal to the square of AC by the converse of Proposition 46. But AB, AC are given, therefore DC is given; and the point C is given, so that D will also be given. But the straight line F is equal to the straight line GH, that is (since GB is twice DB, and BH is twice BC) to twice the straight line DC. Therefore the straight line F is given.

It will be composed as follows.

As twice AB is to AC so let AC be made to CD which is located from the point C towards A; now let the straight line F be taken equal to twice DC. If the point E is taken on AB produced such that CE is less than CB, then the square of DE will be equal to the squares of CE, AD along with the rectangle F, BE. For, by construction, twice the rectangle AB, DC is equal to the square of AC; therefore, by Proposition 46, the square of DB is equal to the squares of AD, BC. Let DG be made equal to DB, and CH to CB, therefore GH is equal to twice DC, that is to the straight line F; hence the rectangle GH, BE, that is the rectangle GE, EB along with the rectangle HE, EB, that is the square of BE along with twice the rectangle DB, BE and the rectangle HE, EB, is equal to the rectangle F, BE. But the square of DB is equal to the squares of AD, BC; therefore the whole quantity, namely the square of DE along with the rectangle HE, EB is equal to the whole quantity, namely the squares of AD, BC along with the rectangle F, BE. Let the rectangle HE, EB be taken away from both sides, and the square of DE will be equal to the squares of AD, CE and the rectangle F, BE [II 5].

Corollary. However if the point E is taken either between the points A, B, or on BA produced on the side of A, then the square of DE along with the rectangle F, BE, that is along with twice the rectangle DC, EB, will be equal to the squares of AD, CE.

A   D   E B   C   E   A   D   B   C
————————————   ————————————

For since it has been shown that the square of DB is equal to the squares of AD, BC, let the square of EC be added to both sides, and the squares of DB, EC will be equal to the squares of AD, BC, EC; let the square of BC be taken away from both sides, and the squares of DB, EB along with twice the rectangle EB, BC will be equal to the squares of AD, EC [II 4]. But the squares of DB, EB are equal to the square of ED along with twice the rectangle EB, BD [II 7]; therefore the square of DE along with twice the rectangles EB, BD and EB, BC, that is along with twice the rectangle DC, EB, will be equal to the squares of AD, EC.

Proposition 48

(This is a Porism to which Propositions 149 and 150 of Pappus's Book 7 apply.)

"Suppose that on a straight line three points A, B, C have been given of which C is on the same side of the point B as A is; on the same straight line a fourth point D will be given which will make the rectangle AE, EC contained by the segments between any point E on AB produced on the side of B and the points A, C, or by the segments between B and the same points A, C, equal to the square of the segment ED between the same point E and the point D which has to be found, along with the rectangle contained by the segment EB between the point E and the remaining given point B, and a certain given straight line."

There are two cases of this Porism; for there are two points D, and two given straight lines which satisfy the same condition.

Case 1. Here the point D which has to be found is on AB produced on the side of B.

Suppose that the Porism is true, and let D be the point and F the straight line which have to be found; and let a point E be taken in any manner on AB produced on the side of B, or let E coincide with B.

$$\overline{\text{A} \qquad \text{C} \qquad \text{B} \qquad \text{D} \qquad \text{E}}$$

First, let the point E be on AD produced beyond D; therefore, by hypothesis, the rectangle AE, EC, that is the rectangle AE, CD along with the rectangle AE, ED, will be equal to the square of DE along with the rectangle F, BE; and when the square of DE has been taken away from both sides, the rectangle AE, CD along with the rectangle AD, DE will be equal to the rectangle F, BE. Again let the point B be taken in place of E, and, by hypothesis, the rectangle AB, BC will be equal to the square of BD; therefore by the first part of Proposition 43 the rectangle AD, DC will be equal to the rectangle contained by both AD, DC and BD. Let the rectangle CD, DE along with the rectangle AD, DE, that is the rectangle which is contained by both AD, DC and DE, be added to both sides, and the whole quantity, namely the rectangle AE, CD along with the rectangle AD, DE will be equal to the whole rectangle, namely that contained by both AD, DC and BE. But it has been shown that the rectangle AE, CD along with the rectangle AD, DE is equal to the rectangle F, BE, which consequently will be equal to the rectangle contained by both AD, DC and BE. Therefore the straight line F is equal to the straight lines AD, DC together. And since the square of BD is equal to the given rectangle AB, BC, the point D will be given; therefore the straight lines AD, DC are given, so that the straight line F is given. These are the things that had to be found.

It will be composed as follows.

A   C   B   D   E

Let the mean proportional BD between AB, BC be found which is located on AB produced on the side of B, and let a straight line F be taken equal to AD, DC together, and let E be any point on AD produced beyond D. The rectangle AE, EC will be equal to the square of DE along with the rectangle F, BE. For since, by construction, the rectangle AB, BC is equal to the square of BD, the rectangle AD, DC will be equal to the rectangle contained by both AD, DC and BD by the first part of Proposition 43. Therefore, as shown in the analysis, the rectangle AE, CD along with the rectangle AD, DE will be equal to the rectangle contained by both AD, DC and BE. Let the square of DE be added to both sides, and the rectangle AE, CD along with the rectangle AE, ED, that is the rectangle AE, EC, will be equal to the square of DE along with the rectangle contained by both AD, DC, that is the straight line F, and BE.

A   C   B   E D

Secondly, let the point E be between the points B, D; therefore since, by hypothesis, the rectangle AE, EC is equal to the square of ED along with the rectangle F, BE, let the rectangle AE, ED be added to both sides, and the rectangle AE, CD will be equal to the rectangle AD, DE along with the rectangle F, BE. Now it will be shown, as in the first analysis, that the rectangle AD, DC is equal to the rectangle contained by both AD, DC and BD, therefore when the rectangle ED, DC has been taken away from both sides, the rectangle AE, CD will be equal to the rectangle AD, DB along with the rectangle CD, BE. Therefore, from the above, the rectangle AD, DE along with the rectangle F, BE is equal to the rectangle AD, DB along with the rectangle CD, BE; let the rectangle AD, DE be taken away from both sides, and the remaining rectangle F, BE will be equal to the remaining part, namely the rectangle AD, BE along with the rectangle CD, BE. Therefore the straight line F is equal to the straight lines AD, DC. Now the point D and the straight lines AD, DC will be shown to be given as in the first analysis; therefore the straight line F is given.

It will be composed as follows.

A   C   B   E D

Let the straight lines BD, F be found as in the first composition; and let the point E be taken in any manner between the points B, D; therefore, as shown in that composition, the rectangle AD, DC will be equal to the rectangle contained by both AD, DC and BD. Now in this second analysis it has been shown that the rectangle AE, CD is equal to the rectangle AD, DB along with the rectangle CD, BE, that is to the rectangles AD, BE, AD, DE and CD, BE; let the rectangle AE, ED be taken away from both sides, and the remaining rectangle AE, EC will be equal to the remaining part, namely the square of ED and the rectangle AD, BE along with the rectangle CD, BE, that is the square of ED and the rectangle contained by both AD, DC and BE, or the rectangle F, BE.

```
     A          C     B        D
     |----------|-----|---------|
                            E
```

Thirdly, let the point E coincide with the point D; therefore, by hypothesis, the rectangle AD, DC, that is, as shown above, the rectangle which is contained by both AD, DC and BD or BE, is equal to the rectangle F, BE; hence also in this case the straight line F is equal to the straight lines AD, DC together. Moreover the point D will be shown to be given as above, and consequently the straight line F will be given.

Now let the same construction be made as was made before, and, as was shown in the preceding compositions, the rectangle AD, DC will be equal to the rectangle contained by both AD, DC and BD, that is the rectangle AE, EC will be equal to the rectangle F, BE.

Corollary 1. Suppose that the point E is taken between the point D and that one of the points C, A which is nearer to the point B, or that it is taken on BA produced on the further side of AC, the rest remaining as before; the rectangle AE, EC along with the rectangle F, BE will be equal to the square of ED.

Case 1. Let the point E be between the points C, B.

```
     A          C  E  B        D
     |----------|--|--|---------|
```

Since, by construction, the rectangle AB, BC is equal to the square of BD, by the second part of Proposition 43 the rectangle contained by both AD, DC and CB will be equal to the square of CD, that is the rectangle contained by both AD, DC and CE along with that contained by both AD, DC and EB will be equal to the square of CD. Let the rectangle DC, CE be taken away from both sides, and the remaining part, namely the rectangle AD, CE along with that contained by both AD, DC and EB will be equal to the rectangle CD, DE. Again let the rectangle CE, ED be taken away from both sides, and the remaining part, namely the rectangle AE, EC along with that contained by both AD, DC and EB, will be equal to the square of ED; that is the rectangle AE, EC along with the rectangle F, BE will be equal to the square of ED.

Case 2. Let the point E be on BA produced on the further side of AC.

E   A   C   B   D

Since the rectangle AB, BC is equal to the square of BD, the rectangle contained by both AD, DC and AB will be equal to the square of AD by the third part of Proposition 43. Let the rectangle contained by both AD, DC and EA be added to both sides, and the whole rectangle contained by both AD, DC and EB will be equal to the whole quantity, namely the rectangle ED, DA and the rectangle EA, CD. Again let the rectangle AE, EC be added to both sides, and the whole quantity, namely the rectangle AE, EC along with that contained by both AD, DC and EB will be equal to the whole quantity, namely the rectangle ED, DA along with the rectangle DE, EA, that is the square of ED. Therefore the rectangle AE, EC along with the rectangle F, BE is equal to the square of ED.

Corollary 2. If the point E is taken between the points A, C, the rest remaining as before, the rectangle AE, EC along with the square of ED will be equal to the rectangle F, BE.

A E   C   B   D

Since, as was shown in Case 1 of the preceding Corollary, the square of CD is equal to the rectangle contained by both AD, DC and CB, let the rectangle contained by both AD, DC and EC be added to both sides, and the whole quantity, namely the rectangle AD, EC along with the rectangle ED, DC, that is the rectangle AE, EC along with the rectangle DE, EC and the rectangle ED, DC, that is the rectangle AE, EC along with the square of ED, will be equal to the whole quantity, namely the rectangle contained by both AD, DC and EB, that is the rectangle F, EB.

A   C   B   D

E

And if the point E coincides with A, the square of AD or ED will be equal to the rectangle F, AB or F, EB. For since the square of CD is equal to the rectangle contained by both AD, DC and CB, let the rectangle contained by both AD, DC and AC be added to both sides, and the whole quantity, namely the rectangle AD, AC along with the rectangle AD, DC, that is the square of AD, will be equal to the whole rectangle contained by AD, DC and AB, that is the rectangle F, AB.

Case 2 of Proposition 48. Here the point D which has to be found is on the same side of the point B as A is.

A   D C   B   E

Let D be the point and F the straight line which have to be found. Therefore if any point E is taken on AB produced on the side of B, by hypothesis, the rectangle AE, EC will be equal to the square of DE along with the rectangle F, BE. Again let the point B be taken in place of E, and, by hypothesis, the rectangle AB, BC will be equal to the square of DB. Therefore, by the first part of Proposition 45 the rectangle contained by the excess of AD, DC and BD is equal to the rectangle AD, DC; and when the rectangle BD, DC has been added to both sides, the rectangle AD, DB will be equal to the rectangle AB, DC. Now the rectangle AE, EC is equal to the square of DE along with the rectangle F, BE, that is the rectangle DE, EC along with the rectangle ED, DC and the rectangle F, BE; thus, when the rectangle DE, EC has been taken away from both sides, the remaining rectangle AD, CE will be equal to the remaining part, namely the rectangle ED, DC and the rectangle F, BE. Let the rectangle AD, DC be added to both sides and the whole rectangle AD, DE will be equal to the whole quantity, namely the rectangle AE, DC and the rectangle F, BE. But it has been shown that the rectangle AD, DB is equal to the rectangle AB, DC; therefore when these have been taken away from the preceding equal quantities, the remaining rectangle AD, BE will be equal to the remaining part, namely the rectangle DC, BE along with the rectangle F, BE; hence the straight line AD is equal to the straight lines DC, F together. And since the square of DB is equal to the given rectangle AB, BC, the point D will be given; hence the straight lines AD, DC and the straight line F will be given. These are the things which have been required to be found.

It will be composed as follows.

Let the mean proportional BD between AB, BC be found which is located on AB on the side of A, and let the straight line F be taken equal to the excess of the straight lines AD, DC; and let a point E be taken in any manner on AB produced beyond B; the rectangle AE, EC will be equal to the square of DE along with the rectangle F, BE.

Since the excess of the straight lines AD, DC is equal to the straight line F, then AD will be equal to the straight line DC along with the straight line F, and the rectangle AD, DE will be equal to the rectangle CD, DE along with the rectangle F, DE. Now, by construction, the rectangle AB, BC is equal to the square of DB; therefore, by the first part of Proposition 45 the rectangle AD, DC is equal to (the rectangle contained by the excess of AD, DC and DB, that is to) the rectangle F, DB; and the rectangle AD, DE has been shown to be equal to the rectangle CD, DE along with the rectangle F, DE; when the preceding equal quantities have been taken away from these, the remaining rectangle AD, CE will be equal to the remaining part, namely the rectangle CD, DE along with the rectangle F, BE. Let the rectangle CE, ED be added to both sides, and the whole rectangle AE, EC will be equal to the whole quantity, namely the square of DE along with the rectangle F, BE.

Corollary 1. Suppose that the point E is taken between the points C, B, or on BA produced on the further side of AC, the rest remaining as before. The rectangle AE, EC along with the rectangle F, BE will be equal to the square of DE.

Case 1. Here the point E is between the points C, B.

$$\text{A} \qquad \text{D} \quad \text{C} \qquad\qquad \text{E} \qquad \text{B}$$

Since the rectangle AB, BC is equal to the square of DB, by the second part of Proposition 45 the rectangle contained by the excess of AD, DC and CB, that is the rectangle F, CB, will be equal to the square of DC; let the rectangle DC, CE along with the rectangle AE, EC be added to both sides, and the whole quantity, namely the rectangle F, CB along with the rectangle DC, CE and the rectangle AE, EC, will be equal to the whole quantity, namely the rectangle ED, DC and the rectangle AE, EC; but the rectangle F, CE along with the rectangle DC, CE is equal to the rectangle AD, CE, because the straight line F along with DC is equal to the straight line AD. Therefore when these equal quantities have been taken away from the preceding quantities, the remaining rectangle F, EB along with the rectangle AE, EC will be equal to the remaining part, namely the rectangle ED, DC and the rectangle DE, EC, that is the square of DE.

Case 2. Here E is on BA produced on the further side of AC.

$$\text{E} \qquad \text{A} \quad \text{D} \quad \text{C} \qquad\qquad\qquad \text{B}$$

Since the rectangle AB, BC is equal to the square of DB, by the third part of Proposition 45 the rectangle contained by the excess of AD, DC and AB, that is the rectangle F, AB, will be equal to the square of AD. Let the rectangle F, EA be added to both sides, and the rectangle F, EB will be equal to the square of AD along with the rectangle contained by the excess of AD, DC and EA. Again let the rectangle EA, DC be added to both sides, and the rectangle EA, DC along with the rectangle F, EB will be equal to the square of AD along with the rectangle EA, AD, that is the rectangle ED, DA. Let the rectangle DE, EA be added to both sides, and the rectangle AE, EC along with the rectangle F, EB will be equal to the square of ED.

Corollary 2. If the point E is taken between the points A, C, the rectangle AE, EC along with the square of DE will be equal to the rectangle F, EB.

Case 1. Here E is between the points C, D.

$$\text{A} \qquad \text{D} \ \text{E} \ \text{C} \qquad\qquad\qquad \text{B}$$

Now it has been shown in Case 1 of Corollary 1 that the square of DC, that is the rectangle DC, CE along with the rectangle CD, DE, is equal to

the rectangle contained by the excess of AD, DC, that is the straight line
F, and CB; moreover the rectangle AD, EC is equal to the rectangle DC, CE
along with the rectangle F, EC, because the straight line AD is equal to the
straight lines DC, F together; when these equal quantities have been added,
the rectangle AC, CE along with the rectangle CD, DE will be equal to the
rectangle DC, CE along with the rectangle F, EB. Let the rectangle DC, CE
be taken away from both sides, and the remaining part, namely the rectangle
AE, EC along with the square of DE will be equal to the rectangle F, EB.

Case 2. Here the point E is between the points A, D.

A E D   C   B

Now it has been shown in Case 1 of Corollary 1 that the square of DC is
equal to the rectangle F, CB, and moreover the rectangle AD, EC is equal to
the rectangle DC, CE along with the rectangle F, EC; when these equal quantities have been added, the square of DC along with the rectangle AD, EC
will be equal to the rectangle DC, CE along with the rectangle F, EB; let the
square of DC along with the rectangle ED, DC, that is the rectangle DC, CE,
be taken away from both sides, and the remaining rectangle AE, EC along
with the square of ED will be equal to the rectangle F, EB.

In the preceding Proposition and its Corollaries the point C is located
between the points A, B; but if the point A is located between the points C,
B, the same demonstrations will serve for this case if A is read for C and C
for A throughout.

These last two Porisms can be resolved without the use of Pappus's Lemmas, but less elegantly.

(p.191)   # Proposition 49

(This is Proposition 155 of Pappus's Book 7.)

"Given a segment of a circle on a straight line AB, to inflect AC, CB in a
given ratio."

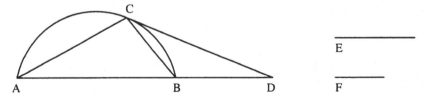

Suppose that this has already been done, and from the point C let the
tangential straight line CD be drawn. "Therefore AD is to DC as CD is to DB,
hence the square of AD is to the square of DC as AD is to DB [Corollary
2 of VI 20]. Now since the angle BCD is equal to the angle BAC in the
opposite segment [III 32], triangles ADC, CDB are equiangular"; therefore

as the square of AC is to the square of CB so (the square of AD is to that of DC, and so) the straight line AD is to the straight line DB. But the ratio of the square of AC to that of CB is given, therefore the ratio of the straight line AD to the straight line DB will also be given; and the two points A, B are given, therefore D is given, "and the tangent DC and the point C will be given; hence AC, BC are also given in position".

Now the Problem will be composed in this way.

Let ACB be indeed a given segment of a circle; moreover let the ratio which E has to F be given. Let AD be made to DB as the square of E is to the square of F, and let the tangent DC be drawn, and let AC, CB be joined. I say that the straight lines AC, CB resolve the Problem.

For AD is to DB as the square of E is to the square of F; moreover the square of AC is to the square of CB as AD is to DB, because CD is tangent to the circle; thus the square of AC will be to that of CB as the square of E is to the square of F. Hence also AC is to CB as E is to F; therefore the inflected lines AC, CB resolve the Problem.

This Proposition, which is corrupted in the Greek text, has been well restored in one or two places by Commandino in his commentary at the letters C, E in the margin; but in that which is found at the letter D he thinks incorrectly that the word δύο has to be deleted from the Greek καὶ 'ἐστι δύο δοθέν, for this indicates that two points A, B are given.

However the Problem can be solved more easily than in Pappus as follows.

Since the base AB of the segment is given, the angle ACB will be given [*Dat*. 92 in the new edition], and, by hypothesis, the ratio AC to CB is given; therefore triangle ACB is given in type [*Dat*. 44], so that the angle ABC is given, and the straight line BC will be given in position. Therefore the point C in which BC meets the circumference, which is given in position, is given.

It will be composed as follows.

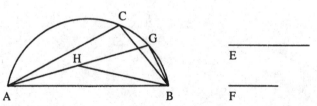

Let AG, BG be drawn to the circumference in any manner, and let HG be made to GB as E is to F, and let BH be joined; moreover let the angle ABC be made equal to the angle GBH, and let AC be joined. Then AC will be to CB as E is to F. For since the angle ABC is equal to the angle GBH, and the angle ACB is equal to the angle AGB in the same segment, the triangle ACB will be equiangular to the triangle HGB; therefore as AC is to CB so (HG is to GB, and so) E is to F.

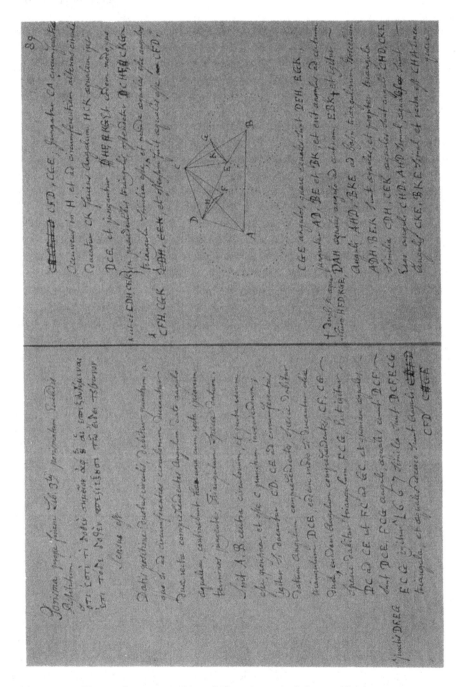

Simson's restoration of the antepenultimate Porism
Adversaria, Vol. I, pp. 88–89
(Courtesy of Glasgow University library)

This problem is of use in the construction of the following Porism, which indeed appears to be the antepenultimate of Euclid's Book 3, and which is stated rather briefly by Pappus in the following words, viz.

῞Οτι ἔστι τι δοθὲν σημεῖον ἀφ' οὗ αἱ ἐπιζευγνύμεναι ἐπὶ τόδε * δοθὲν περιέξουσι τῶι εἴδει τρίγωνον.

This statement, which is certainly mutilated, and perhaps corrupted, seems to require to be expanded and explained as in the following Proposition.

191)

Proposition 50

"Suppose that two circles have been given in position; a point will be given with the property that, if two straight lines containing an angle equal to a given angle are drawn from it to the circumferences of the circles, then these straight lines along with the straight line joining their extremities will contain a triangle which is given in type."

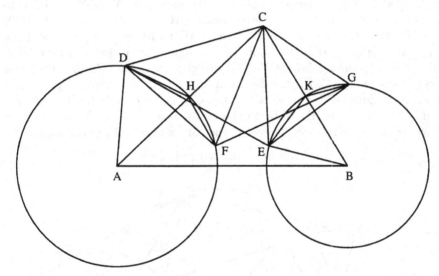

Let A, B be the centres of the circles, and suppose that C is the point which has to be found; therefore if CD, CE which contain the given angle are drawn to the circumference, either the convex part or the concave part (n1), and DE is joined, the triangle DCE will be given in type. Likewise, if another two straight lines CF, CG, containing an angle equal to the angle DCE, are drawn and FG is joined, the triangle FCG will be given in type. Therefore since triangles DCE, FCG are similar, FC is to CG as DC is to CE, and *permutando* ⟨DC is to CF as EC is to CG⟩; and since the angles DCE, FCG are equal, the angles DCF, ECG will also be equal, and the sides

---

* Perhaps τόνδε sc. κύκλον.

about them are proportional. Therefore the triangle DCF is equiangular to the triangle ECG [VI 6], and consequently the angle CFD is equal to the angle CGE. Let AC be joined, and let it meet the circumference in H; now let CK be drawn to the circumference of the circle whose centre is B in such a way that the angle HCK is equal to the angle DCE, and let FH, GK, DH, EK be joined; therefore, as in the above, it will be shown that the triangle CHF is similar to the triangle CKG, and also that the triangle CDH is similar to the triangle CEK; hence the angles CFH, CGK are equal, and the angles CFD, CGE have been shown to be equal; therefore the angles DFH, EGK are also equal. Let AD, BE, BK be joined; the angles at the centres DAH, EBK, which are in fact twice the equal angles DFH, EGK, will be equal to each other. Therefore the angles AHD, BKE at the bases of the isoceles triangles ADH, BEK are equal; and since the triangles CDH, CEK are similar, the angles CHD, CKE are equal; therefore the angles AHD, CHD together are equal to the angles BKE, CKE together; but AHC is a straight line, therefore BKC is also a straight line [I 14]. And since in the triangles ADC, BEC the angles DAC, EBC are equal, as also are the angles DCA, ECB, as AC is to AD so BC will be to BE, and *permutando* ⟨AC is to BC as AD is to BE⟩; but AD, BE are given, so that the ratio AC to BC is also given; and the angle ACB is given (n2), therefore the triangle ABC is given in type [*Dat.* 44]. And since the angles BAC, ABC are given and the straight line AB is given in position, AC, BC will be given in position; therefore the point C is given. Now since DC is to CE as AC is to CB, and the angles DCE, ACB are equal, the triangle DCE will be similar to the triangle ACB, which has been shown to be given in type. These are the things that had to be shown.

It will be composed as follows.

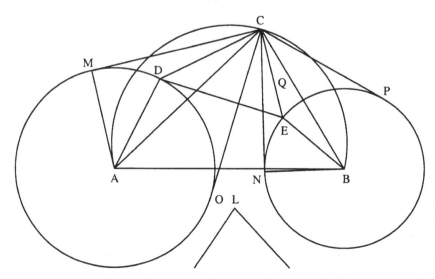

Let A, B be the centres of the circles which are given in position, and let the angle at L be given; let AB be joined and above it let the circular segment be described for which the angle is equal to the angle at L, and, by means of the preceding Problem, let AC, BC be drawn to the circumference of the segment so that they have the given ratio which the radius of the circle centre A has to the radius of the circle centre B; then C will be the point which has to be found. That is to say, if CD is drawn in any manner from the point C to the circumference whose centre is A, and another straight line CQ is drawn from the same point making an angle with CD equal to the angle ACB, or the given angle at L, then CQ will meet the circumference whose centre is B in a certain point E which will make the triangle DCE similar to the triangle ACB.

For from the point C let straight lines CM, CN be drawn tangent to the circles on one side and let CO, CP be drawn tangent to them on the other side, and let AM, BN be joined. Therefore since, by construction, AC is to BC as AM, which is a radius from the centre A, is to BN, which is a radius from the centre B, the right-angled triangles AMC, BNC will be equiangular [VI 7]; therefore the angles ACM, BCN are equal, and consequently the angles MCN, ACB are equal. It will be shown similarly that the angles OCP, ACB are equal; therefore the angles MCN, OCP are equal. Therefore since the straight line CD lies within the angle MCO, the straight line CQ which makes the angle DCQ equal to the angle ACB, that is to the angle MCN, or to OCP, will necessarily lie within the angle NCP, so that it will necessarily meet the circumference also. Let it meet the circumference in E which is on the same side of the circumference as D, either the convex or the concave side, and let AD, BE be joined. Therefore since the angles ACB, DCE are equal, the angles ACD, BCE will be equal, and, by construction, AC is to CB as AD is to BE, and *permutando*, AC is to AD as CB is to BE, and, because of the circle, the remaining angles ADC, BEC are both either greater than or less than a right angle, therefore (n3) the triangles ADC, BEC will be equiangular [VI 7]. Therefore BC is to CE as AC is to CD, and *permutando* [AC is to CB as DC is to CE]; and since the angles ACB, DCE are equal, the triangle DCE will be similar to the triangle ACB.

Corollary. And by means of this the following Porism will be solved; it was proposed to me as a Problem by Mr William Trail, a distinguished young man, who has made great progress in Mathematics for his years.

Porism. "Suppose that two circles DHF, EKG have been given in position, whose centres are A, B; a point will be given with the property that, if two straight lines containing an angle equal to a given angle are drawn from it to the circumferences, these along with the straight line which joins their extremities, one of which is on the concave circumference, the other on the convex circumference, will contain a triangle which is given in magnitude."

Suppose that C is the point which has to be found; therefore if CR, CE are drawn to the circumferences, making the given angle, and RE is joined, by hypothesis, the triangle CRE will be given in magnitude (n4). And since the angle RCE of the triangle CRE is given, the rectangle RC, CE will have a given ratio to the triangle CRE by the Corollary to Proposition 62 of the *Data* which were published in Glasgow, and the triangle is given in magnitude, so that the rectangle RC, CE is given. Similarly, if another two straight lines CS, CG are drawn (making the given angle), it will be shown that the rectangle SC, CG is given, which consequently is equal to the rectangle RC, CE; therefore SC is to CR as EC is to CG. Let RC meet the circumference again in D, and let SC meet it again in F, and because of the circle SC will be to CR as CD is to CF; hence EC is to CG as DC is to CF. Now it will be shown, as in this Proposition, that AC is to CB as AD is to BE, and that the point C is given.

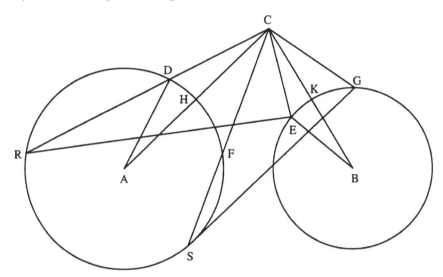

It will be composed as follows.

Let the point C be found as in this Proposition, and let AC meet the concave circumference in T, and let BC meet the convex circumference in K; if from the point C two straight lines CR, CE are drawn, one to the concave circumference, the other to the convex circumference in any manner (so that the angle RCE is equal to the given angle), and RE is joined, as also TK, then the triangle RCE will be equal to the given triangle TCK. (For let RC meet the circumference again in D, and) let TC meet the circumference again in H, and, by this Proposition, the triangles DCE, HCK will be similar; therefore EC is to CK as DC is to CH; but as DC is to CH so on account of the circle is TC to CR, therefore TC is to CR as EC is to CK; hence the rectangle RC, CE is equal to the rectangle TC, CK. Now since the angles RCE, TCK are equal, the rectangle TC, CK will be to the triangle TCK as the rectangle

RC, CE is to the triangle RCE; but the rectangles have been shown to be equal, therefore the triangle RCE is equal to the given triangle TCK. Q.E.D.

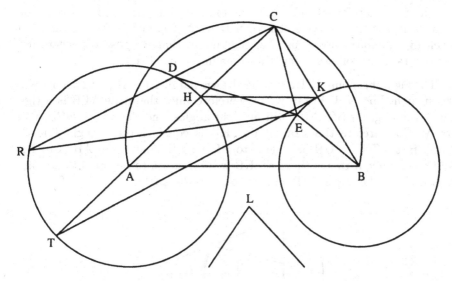

5th January, 1767.

92)

Proposition 51

(This is Proposition 156 of Pappus's Book 7 and his Lemma 30.)

"Let there be a circle whose diameter is AB, and from any point D on the circumference let DE be drawn perpendicular to the diameter, and let DF be drawn in any manner to meet the circumference again in F, and AB in H; now let the join EF be produced to meet the diameter in G. I say that AH is to HB as AG is to GB."

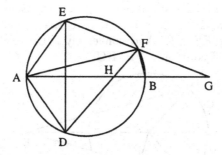

For let DA, AE, AF be joined; and since DE is perpendicular to the diameter, the angle DAB will be equal to the angle BAE [III 3 and I 4]; but the angle DAB is in fact equal to the angle HFB in the same segment, while the angle BAE is equal to the angle BFG, which is an exterior angle of the cyclic quadrilateral AEFB [III 22]. Therefore the angle HFB is equal to the angle BFG; and the angle AFB in the semicircle is a right angle. Therefore, by the Lemma, AH is to HB as AG is to GB.

The Lemma referred to is not to be found in Pappus, but it is the Proposition which is the converse of Proposition 52 of Pappus's Book 6, and it is given by Commandino in his Commentary to that Proposition, where it is demonstrated as follows, a few things having been changed.

Through the point B let KL be drawn parallel to AF, and let it meet the straight lines FH, FG in K, L; therefore since the angle AFB is a right angle, the angles FBK, FBL will be right angles, and the angles KFB, BFL are equal, therefore the straight line KB is equal to the straight line BL [I 26]. But AG is to GB as (AF is to BL, or KB, that is as) AH is to HB. And conversely, if this is the case, KB will be shown to be equal to BL, and consequently the angle KFB will be equal to the angle BFG.

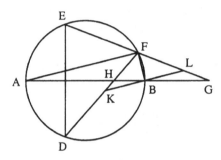

Corollary. Hence if the point H is given, the rest remaining as in the Proposition, the point G will be given, and conversely; the circle has of course been given in position.

(p.192)   **Proposition 52**

"If straight lines EFH, EKG are drawn from a point E to a circle BCD and cut off equal arcs FH, KG, the drawn lines EH, EG will be equal to each other, as also will be EF, EK. And conversely if EH, EG, or EF, EK, are equal, they will cut off equal arcs FH, KG."

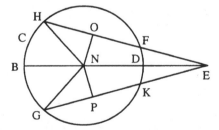

First, let the arcs FH, KG be equal; therefore the straight line FH will be equal to the straight line KG; hence the perpendiculars NO, NP drawn from the centre N to these straight lines will be equal [III 14]. Therefore the straight line EO will be equal to the straight line EP [I 47]; and OH, PG are equal; therefore EH, EG, as also EF, EK, are equal to each other.

And if EH, EG are equal, they will cut off equal arcs of the circumference FH, KG. For if straight lines NH, NG are drawn from the centre, in the triangles HNE, GNE, the angles NEH, NEG will be equal [I 8]; hence in the right-angled triangles ENO, ENP the straight lines NO, NP will be equal [I 26]. Therefore the straight lines FH, GK will be equal [III 14], and the arcs which they cut off will be equal.

These two Propositions are Lemmas to the penultimate Porism of Book 3 of Euclid's Porisms; the Porism is expressed in the following words in the Greek text of the Preface of Pappus of Alexandria to his Book 7; the most distinguished Halley prefixed the Preface to his edition of Apollonius's *The Cutting off of a Ratio*, which was printed at Oxford in the year 1706:

῞Οτι ἔστι τι δοθὲν σημεῖον ἀφ' οὗ αἱ ἐπιζευγνύμεναι ἐπὶ τόδε ἴσας ἀπολαμβάνουσι περιφερείας.

I have found, with no little investigation, that the Porism described briefly and rather obscurely in these words has to be explained as follows, namely:

192)
Proposition 53

"Suppose that a point is given inside or outside a circle which is given in position, and let a straight line meeting the circle be drawn through it in any manner; another point will be given with the property that, if two straight lines are drawn from it to the points in which the drawn line meets the circumference, these will cut off equal arcs from the circle."

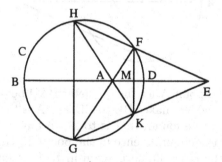

Suppose that the Porism is true; and let the point A be given inside or outside the circle BCD which is given in position; and let E be the point which has to be found, namely the point such that, if any straight line FAG is drawn through A to the circle and the joins EF, EG meet the circumference again in H, K, then the arcs FH, KG are equal to each other. Let KA be joined, and since KA passes through the given point A, and EKG has been drawn cutting off the arc KG, it is necessary, by hypothesis, that the straight line drawn from E to the point in which KA meets the circumference again will cut off an arc equal to the arc KG, namely the arc FH. Now it is clear from [III 8] and [III 29] that no straight line apart from the straight line EFH can cut off an arc on the side C of the circle which is equal to the arc FH; therefore the straight line KA passes through the point H. Now since the arcs FH, GK are equal to each other, if FK, HG are joined, the angle HGF will be equal to the angle GFK; and they are alternate angles, so that FK, HG are parallel. Therefore GA is to AF as (GH is to KF, that is as) GE is to EK. Now since the straight lines EFH, EKG cut off equal arcs, the straight line EK will be equal to the straight line EF by Proposition 52. Therefore GE is to EF as GA is to AF; hence the join EA bisects the angle FEK [VI 3]. Let FK meet AE in M, and since FE, EM are equal to KE, EM, and they contain equal angles, the straight line EM will bisect the straight line FK at right angles [I 4], as also GH, so that it will also pass through the centre of the circle. Let EA meet the circumference in the points B, D; and since HG is perpendicular to the diameter BD, and the straight lines GAF, HFE have been drawn, BA will be to AD as BE is to ED by Proposition 51; but BA, AD are given, so that the ratio BE to ED is given; and BD is given in magnitude, so that DE is also given in magnitude; and it is also given in position, for the centre of the circle is given, and the point A is given on the straight line BDE; therefore the point E is given. This is what had to be found.

It will be composed as follows.

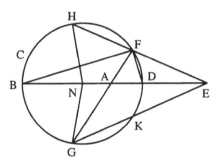

Through the given point A let the diameter BAD be drawn, and on it let the point E be taken which makes BE to ED as BA is to AD; then E will be the point which has to be found; that is to say, if any straight line is drawn through A, meeting the circumference in the points F, G, and if EF, EG are joined which meet the circumference again in H, K, then the arcs FH, GK will be equal to each other.

For let BF, FD be joined; therefore the angle BFD in the semicircle is a right angle; and since BA is to AD as BE is to ED, the angles AFD, DFE will be equal to each other by the converse of the Lemma at the end of Proposition 51; therefore the angles DFE, BFA are together equal to the right angle BFD, and the angles DFE, BFH are together equal to a right angle; therefore the angles BFA, BFH are equal; hence, if NH, NG are drawn from the centre N, the angles at the centre BNH, BNG will be equal, and consequently the angles HNE, GNE, which are supplementary to them, are equal to each other. Therefore in the triangles HNE, GNE the base HE is equal to the base GE [I 4]; therefore the straight lines EH, EG cut off equal arcs FH, KG by Proposition 52.

Corollary 1. From a point E let two straight lines EFH, EKG be drawn to the circle BCD, cutting off equal arcs FH, KG, and let the join FG, or HK, meet the diameter BD which passes through E in A; then BE will be to ED as BA is to AD. And if the circle is given in position, and one of the points A, E is given, say E, then the other point A will be given; and consequently FG will pass through a given point.

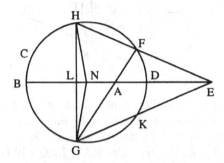

For let GH be joined, which BE meets in L, and let N be the centre, and let NG, NH be joined. Therefore since EH, EG cut off equal arcs, EH will be equal to EG by Proposition 52, therefore in the triangles HEN, GEN the angles NEH, NEG are equal [I 8]; hence in the triangles LEH, LEG the angles HLE, GLE are equal to right angles [I 4]. Therefore since HG is perpendicular to the diameter BD, and GAF, HFE have been drawn, BE will be to ED as BA is to AD by Proposition 51. And if the point E is given, A will also be given, and conversely, the circle having been given in position.

Corollary 2. On the diameter BD of a circle let any point A other than the centre be taken, and on the same diameter let the point E be taken such that BA is to AD as BE is to ED; and from the points A, E let AF, EF be inflected to the circumference in any manner, meeting it again in the points G, H; the join GH will be perpendicular to the diameter BD.

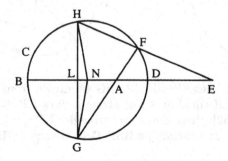

For since the angles LNH, LNG have been shown to be equal in the
preceding composition (n), the angles at L in the triangles NLH, NLG will
be right angles. Therefore GH is perpendicular to LN.

Corollary 3. Suppose that the same conditions hold as in Corollary 2.
In addition let EG be joined, and let it meet the circumference again in K;
the points H, A, K will be in a straight line.

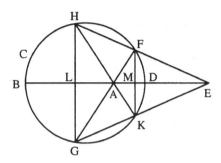

For let FK be joined; therefore since the arcs FH, GK are equal by the
composition of this Proposition, the straight lines HG, FK will be parallel,
and HG is perpendicular to BD, and so also to FK. And since LM bisects
the straight lines HG, FK, the triangles HAL, LAG will be equal, as also the
triangles FAM, MAK; therefore the angle HAL is equal (to the angle LAG,
that is to the angle FAM, that is) to the angle MAK. Therefore the points
H, A, K are in a straight line.

(p.193) # Proposition 54

"If from a point K two straight lines KMP, KQL are drawn to a circle DCE
cutting off equal arcs MP, QL, and equal arcs PH, LG are taken from the
arcs PM, LQ, then the joins KH, KG will cut off equal arcs."

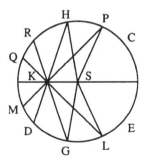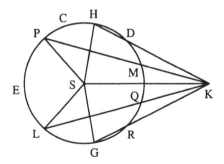

For let straight lines SP, SH, SL, SG be drawn from the centre S; and
since KP, KL cut off equal arcs, the straight lines KP, KL will be equal to
each other [Prop. 52]; therefore in the triangles KPS, KLS the angles KSP,
KSL are equal [I 8]; and the angles HSP, GSL are equal [III 27], therefore the

remaining angles KSH, KSG are equal; therefore the base KH is equal to the base KG [I 4]; hence the straight lines KH, KG cut off equal arcs, viz. HD, GR [Prop. 52].

Proposition 55

"Suppose that from the two points A, B straight lines AC, BC are inflected to the circumference of the circle CDE, meeting it again in the points D, E, and from one of them E the straight line EF is drawn parallel to the join AB to meet the circumference again in F, and the join DF meets the straight line AB in G; then the rectangle BA, AG will be equal to the rectangle CA, AD. And conversely, if the point G is taken on the straight line AB which makes the rectangle BA, AG equal to the rectangle CA, AD, and the join GD meets the circumference again in F, then the join FE will be parallel to the straight line AB."

Since, in the first part, AB, FE are parallel, the angle AGF is equal to the angle DFE, that is to the angle DCE in the same segment; and in the triangles ACB, AGD either the angle BAC is common or it is equal to the angle DAG. Therefore BA is to AC as AD is to AG [VI 4], and consequently the rectangle BA, AG is equal to the rectangle CA, AD.

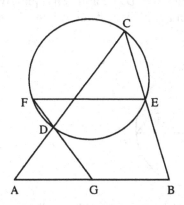 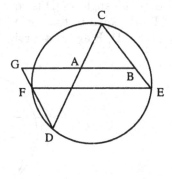

And since, in the second part, the rectangle BA, AG is put equal to the rectangle CA, AD, then BA will be to AC as AD is to AG; therefore the angle AGD is equal to the angle ACB or DCE [VI 6], that is to the angle DFE in the same segment. Hence the straight line FE is parallel to the straight line AB.

Corollary. And in the case where AB is outside the circle, if the rectangle BA, AG is equal to the rectangle CA, AD, and the straight line EF is parallel to the straight line AB, the points G, D, F will be in a straight line.

For let DH be drawn parallel to AB; therefore the angle AGD will be equal to the angle GDH. Now since the rectangle BA, AG is equal to the rectangle CA, AD, the angle AGD will be equal to the angle ACB or DCE, that is to the angle DFE. Therefore the angle GDH is equal to the angle DFE. But the angle DFE along with the angle FDH is equal to two right angles, because DH, FE are parallel; hence the angle GDH along with the angle FDH is also equal to two right angles, and consequently the points G, D, F are in a straight line [I 14].

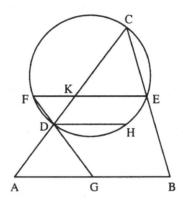

Or rather, it may be shown thus. Let AD meet the straight line FE in K; therefore the angle GAD is equal to the angle DKF, and, as before, it will be shown that the angle AGD is equal to the angle DFK; and consequently the angle ADG is equal to the angle FDK, and ADK is a straight line, therefore the points G, D, F are in a straight line.

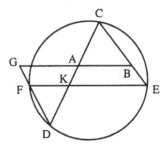

In the case where AB is inside the circle the same thing will be shown thus.

Let DC meet the straight line FE in K; now BA is to AC as EK is to KC; but as BA is to AC so AD is to AG by hypothesis, and as EK is to KC so, because of the circle, KD is to KF; therefore KD is to KF as AD is to AG, and AG, KF are parallel, therefore the points D, F, G are in a straight line.

Proposition 56

"Let ABCD be a quadrilateral whose two adjacent sides AB, AD are equal to each other, as also are the other two sides CB, CD; if BE, DE are drawn making the angles CBE, CDE equal, the points A, E, C will be in a straight line."

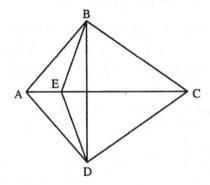

For let AE, CE, BD be joined, and since CB, CD are equal, the angles CBD, CDB will be equal, so that the angles EBD, EDB are also equal; therefore the straight line BE is equal to the straight line ED. And since CE, EB are equal to CE, ED and the base CB is equal to the base CD, the angle CEB will be equal to the angle CED. Again since AE, EB are equal to AE, ED, and the base AB is equal to the base AD, the angle AEB will be equal to the angle AED. Therefore the angles AEB, BEC are together equal to the angles AED, DEC together; and all four make up four right angles, so that half of them, namely the angle AEB along with the angle BEC, is equal to two right angles; therefore the points A, E, C are in a straight line.

The three preceding Propositions are Lemmas for the final Porism of Book 3 of Euclid's Porisms, which is enunciated in the Greek text of Pappus's preface to his Book 7 in the following words.

῞Οτι ἥδε ἤτοι ἐν παραθέσι ἔσται, ἢ μετά τινος εὐθείας ἐπὶ τὸ δοθὲν "σημεῖον" νευούσης δοθεῖσαν περιέχει γωνίαν.

That this straight line either will be parallel to a straight line which is given in position, or will contain a given angle with a certain straight line going to a given "point".

After I had shown the Porism in Proposition 49 to Mr Matthew Stewart, I asked him to attempt an explanation of this, which I said I believed had to be interpreted as being about two straight lines inflected from given points to the circumference of a circle which is given in position. After a short time he wrote to me that he had found this to be true, and indicated the construction of the Porism. But the Porism can be enunciated more explicitly and generally as follows. See the figures for Proposition 57.

"Suppose that from two given points A, B two straight lines AC, BC are inflected in any manner to the circle CDE which is given in position and they meet the circumference again in D, E; the join DE will contain a given angle with a straight line going to a given point, or it will be parallel to a straight line which is given in position, or it will go to a given point."

However this general Porism is divided into another three, whose investigation I have carried out in the following Propositions 57, 58, 61.

(p.193)

Proposition 57

(This is the first part of the preceding Porism.)

"Here the straight line which joins the points of intersection of the inflected lines and the circumference contains a given angle with a straight line which goes to a given point."

Case 1. Here the points A, B are on a straight line which passes through the centre of the circle.

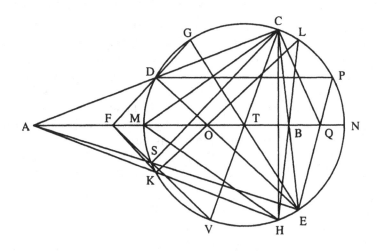

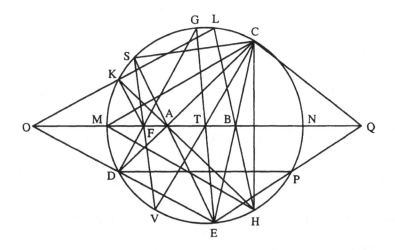

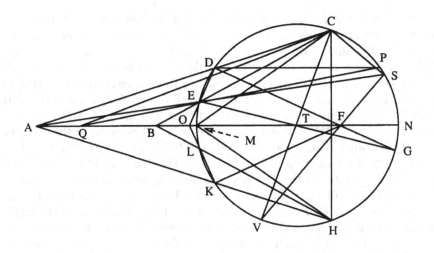

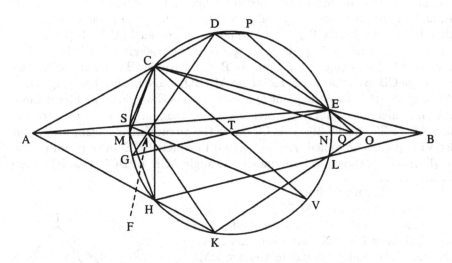

Let the points A, B be given, and let the circle CDE be given in position; and let straight lines AC, BC be inflected in any manner to the circumference, meeting it again in D, E, and let DE be joined; suppose that the Porism is true, so that the straight line DE contains a given angle with a straight line which goes to a given point, and this angle is either at the intersection of the straight line DE and the straight line AC, or at the intersection of the straight line DE and the straight line BC; suppose that it is at the point of intersection D of ED and AC; and suppose that F is the point which has to be found, and DF having been joined let it be produced to meet the circumference again in the point G; therefore, by hypothesis, the angle EDF is given, as also is the angle which is supplementary to it. From the point C let CH be drawn perpendicular to AB, and to the point H in which it meets the circumference again, let AH, BH be inflected which meet the circumference again in K, L, and let LK, KF be joined; therefore, again by hypothesis, the angle LKF is given which consequently is equal to the angle EDF. Let AB meet the circumference in M, N, and let ED, LK meet each other in O; and since CH is perpendicular to the diameter MN, the straight line AC will be equal to the straight line AH [I 4], therefore the arcs DC, KH which they cut off are equal to each other [Prop. 52], and CM, MH are equal, therefore the arcs DM, MK are also equal (n1). Likewise since the straight lines BC, BH are equal, the arcs LN, NE will be shown to be equal; therefore O, the point of intersection of ED, LK, is on the diameter MN. * And since the arc DM is equal to the arc MK, the straight line OD will be equal to the straight line OK, and the straight line AD is equal to the straight line AK, and the angle EDF is equal to the angle LKF, so that also the angle ODF is equal to the angle OKF; therefore the point F is on the straight line AO or AB [Prop. 56]. Let DP be drawn parallel to AB, and let it meet the circumference again in P, and let the join EP meet AB in Q; and since AC, BC have been inflected to the circumference at C and meet it again in D, E, and DP has been drawn parallel to AB, and the join EP meets AB in Q, then the rectangle AB, BQ will be equal to the rectangle CB, BE [Prop. 55]; therefore the triangle BQE is equiangular to the triangle ACB (n2), so that the angle BQE is equal to the angle ACB; but the point B is given, so that the rectangle CB, BE, that is the rectangle AB, BQ, is given, and the straight line AB is given, therefore the point Q is given. Let AE be joined and let it meet the circumference again in S, furthermore let the join EG meet the straight line AB in T, and let the join CT meet the circumference again in V, and let VF, QC, CS be joined. And since (n3) the angle BQE, or in certain cases TQE, has been shown to be equal (to the angle ACB, or DCE, that is) to the

---

* This is shown as follows.

Since the arcs MLN, MEN are equal, when the equal arcs MK, MD have been added and the equal arcs LN, EN have been taken away, the arc KML will be equal to the arc DME, therefore the straight lines KL, DE are equal; let perpendiculars TR, TS be drawn to them from the centre T, and, TO having been joined, since TR, TS are equal [III 14], OR, OS will also be equal [I 47], so that OL, OE are also equal; and TL, TE having been joined, the an-

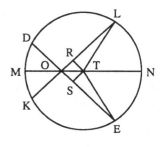

gles OTL, OTE will be equal, and the angles LTN, ETN are equal, therefore the angles OTL, LTN together are equal to the angles OTE, ETN together; and all four angles are equal to four right angles, therefore the angles OTL, LTN, which are half of them, are together equal to two right angles, and consequently the points O, T, N are in a straight line, and the point O will be on the diameter TN.

angle DGE in the same segment, and the angle ETQ is equal to the angle FTG, the triangle QTE will be equiangular to the triangle GTF; therefore the rectangle QT, TF is equal to the rectangle (GT, TE, that is [III 35] to the rectangle) CT, TV; therefore the angle CQT is equal to the angle TVF (n4). Now since the rectangle AB, BQ has been shown to be equal to the rectangle CB, BE, the triangle QBC will be equiangular to the triangle EBA, so that the angle BQC, or CQT, is equal to the angle BEA, or CEA; but it has been shown that the angle CQT is equal to the angle TVF; therefore (n5) the angle BEA, or in certain cases CES, is equal to the angle TVF, so that they stand upon equal arcs; but the angle BEA, or CES, stands upon the arc CS, and the angle TVF, or CVF, has a common extremity C with the angle CES; therefore the angle CVF stands upon the arc CS, that is the straight line VF passes through the point S. Now since AE, BE have been inflected from the points A, B and meet the circumference again in S, C, the angle CSF, or CSV, is equal, by hypothesis, to the angle EDF, or EDG, and the angles which are supplementary to them are equal to each other, therefore the straight line CV is equal to the straight line EG. Therefore since the equal straight lines CV, EG pass through the same point T on the diameter, and they make unequal angles with it, for CT, HT, and not CT, ET, make equal angles with the diameter, consequently T will be the centre of the circle. *
And since the rectangle QT, TF is equal to the rectangle GT, TE, that is to the square of the semidiameter, the rectangle QT, TF will be given, and QT is given, so that TF is also given; and the point T is given, so that F is given; and since EG is a diameter of the circle, for it passes through the centre T, the angle EDG, or EDF, will be a right angle. Therefore the straight line ED contains a given angle EDF with the straight line DF which goes to the given point F. This is what had to be shown.

---

* This is shown as follows.

If two equal straight lines in a circle pass through the same point on a diameter which is not the centre, they will make equal angles with the diameter, as can be easily shown by drawing perpendiculars from the centre to the equal straight lines. And consequently, if two equal straight lines which pass through the same point on the diameter make unequal angles with it, the point through which they pass will be the centre of the circle.

It will be composed as follows.

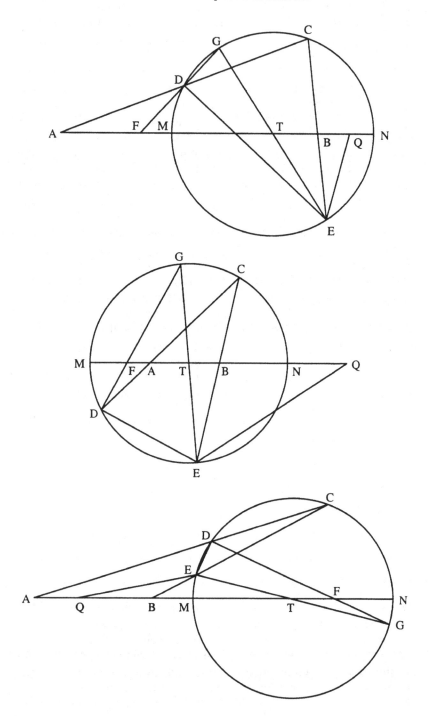

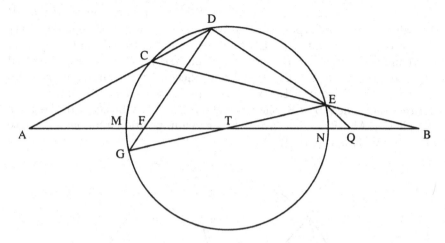

Let A, B be two given points on a diameter, and let the circle CDE be given in position; let AB meet the circumference in M, N, and let the rectangle AB, BQ be made equal to the rectangle NB, BM, and let the point Q be located so that the points A, Q are on the same side or on opposite sides of the point B, according as the points M, N are on the same side or on opposite sides of the same point B. Let T be the centre of the circle which, by hypothesis, is on the straight line AB, and let the square of MT be made equal to the rectangle QT, TF, and let the point F be located on the opposite side of the centre T from Q. If from the points A, B straight lines AC, BC are inflected to the circumference in any manner and meet it again in D, E, and ED, DF are joined, the angle EDF will be a right angle. For let the join ET meet the circumference again in G, and let GF, QE be joined. And since the rectangle QT, TF is equal (to the square of MT, that is) to the rectangle GT, TE, the triangle QTE will be equiangular to the triangle GTF, so that the angle TQE, or BQE, is equal to the angle TGF. And since the rectangle AB, BQ is equal (to the rectangle MB, BN, that is) to the rectangle CB, BE, the triangle ABC is equiangular to the triangle EBQ, so that the angle BCA, or ECD, is equal to the angle BQE, or, in certain cases, to the angle TQE, that is, as has been shown, to the angle TGF, or EGF. Therefore the angles ECD, EGF stand upon equal arcs, but the angle ECD stands upon the arc ED, and they have a common extremity E, therefore the angle EGF stands upon the same arc ED, that is the straight line GF passes through the point D, in other words the points G, F, D are in a straight line. Therefore since EG is a diameter, the angle EDG, or EDF, in the semicircle is a right angle. Q.E.D.

Case 2. Here the given points A, B are not on a straight line which passes through the centre of the circle, but the rest remains as before.

Let straight lines AC, BC be inflected in any manner to the circumference from the given points A, B, and let them meet the circumference again in D, E; let F be the point which has to be found, and let the join DF meet the circumference again in G. Therefore, ED having been joined, the angle EDF will be given by hypothesis, as also the angle which is supplementary to it. Through the point E let EH be drawn parallel to AB, and let it meet the circumference again in H, and let the join DH meet the straight line AB in K. Therefore since AC, BC have been inflected to the circumference and meet it again in D, E, and EH has been drawn parallel to AB, and the join DH meets the same AB in K, the rectangle BA, AK will be equal to the given rectangle CA, AD [Prop. 55]. And AB is given in position and magnitude, so that the point K is given.

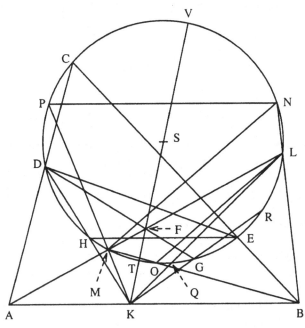

Let AF be joined and let it meet the circumference in the points L, M, and let the join BL meet the circumference again in N, while the join BM meets it again in O; also let KM be joined and let it meet the circumference again in P, and let PN be joined. Therefore since AL, BL have been inflected to the circumference from the points A, B and meet it again in M, N, and the rectangle BA, AK is equal (to the rectangle CA, AD, that is) to the rectangle LA, AM, and the join KM meets the circumference again in P, the straight line PN will be parallel to the straight line AB [Prop. 55], and consequently to the straight line HE; therefore the arc HP is equal to the arc EN. Let MN, LO be joined, and since AL, BL have been inflected from the points A, B and meet the circumference again in M, N, by hypothesis, the straight line NM will contain given angles with the straight line ML which passes through the given point F; and similarly since AM, BM have been inflected from the same points A, B and meet the circumference again in L, O, the straight line OL will contain given angles with the straight line LM which passes through the point F. Therefore since the angle EDF is given, and the angle NMF is equal to the same given angle, as also is the angle OLF, the three arcs GE, LN, MO, on which they stand, will be equal to each other.

But in the case where the straight line AB meets the circle, if the point D, or M, or L is on the smaller segment cut off by the straight line AB, in place of the angle EDF, or NMF, or OLF, the angle which is supplementary to it has to be taken, FD, or FM, having been produced (n6); and in these cases also the arcs GE, LN, MO, namely those on which the angles stand which are equal to those which are supplementary, will be equal by [III 22].

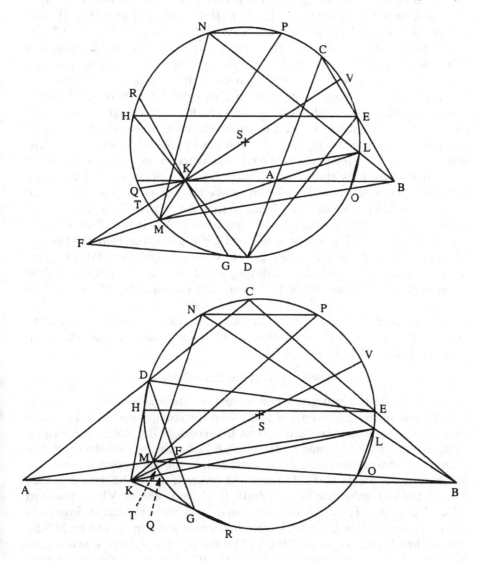

Moreover since the rectangle BA, AK is equal to the rectangle LA, AM, the points B, K, L, M will be on a circle; let KL be joined and let it meet the circumference again in Q; therefore the angle KMB will be equal to the angle KLB, if the points M, L are on the same side of the straight line AB; but if they are on opposite sides of it, the angle KMB will be equal to the angle QLN which is supplementary to the angle KLB; therefore in both cases the angles BMP, QLN are equal. Therefore the arc OMP in which the angle BMP, or OMP, lies is equal to the arc QLN in which the angle QLN lies. And it has been shown that the arc MO is equal to the arc LN, therefore also the arc MP is equal to the arc QL. Let the join KG meet the circumference again in R, and since the straight lines KMP, KQL cut off equal arcs MP, QL, and the arc HP is equal (to the arc EN, that is, because the arcs GE, LN are equal) to the arc GL, the straight lines KHD, KGR will cut off equal arcs HD, GR [Prop. 54]; therefore the straight line GD goes to a given point on the diameter which passes through the point K, by Corollary 1 of Proposition 53. But DG goes through the given point F by hypothesis; therefore the point F is on the diameter which passes through K, for only a single point can be given on the straight line DG; for if two points were given on it, the straight line DG would be given in position, which is impossible, since the straight line ADC has been inflected to the circumference in any manner; therefore the join KF will pass through the centre S of the circle. Let KS meet the circumference in the points T, V, and since the points K, S are given, the straight line KTV is given in position, so that the straight lines KT, KV are given. And since by Corollary 1 of Proposition 53 VK is to KT as VF is to FT, the ratio VF to FT will be given; and TV is given, therefore the point F is given. But the point A is given, therefore the straight line AMFL is given in position, and the points M, L will be given; hence the straight line BLN is given in position, for the point B is given; and consequently MN is given in position. Therefore the angle NML, and the angle EDG which is equal to it or is supplementary to it, are given. Therefore the straight line ED contains the given angle EDF with the straight line DF which passes through the given point F. This is what had to be found.

It will be composed as follows.

Let A, B be the two given points, and let the circle CDE be given in position and magnitude; from the point A let the straight line ADC be drawn to the circle in any manner, and (n7) let the rectangle BA, AK be made equal to the rectangle CA, AD, and let the point K be located so that the point A is outside or between the points K, B, according as the same point A is outside or between the points C, D. Through the centre S let KS be drawn and let it meet the circumference in the points T, V, and (n8) let VF be made to FT as VK is to KT; let AF be joined, and let it meet the circumference in L, M, and let the join BL meet the circumference again in N, and let MN be joined; then F and the angle NML will be the point and angle which had to be found; that is to say, if straight lines AC, BC are inflected from the points

A, B to the circumference in any manner and they meet it again in the points D, E, and ED, DF are joined, the angle EDF, or its supplementary angle, will be equal to the given angle NML.

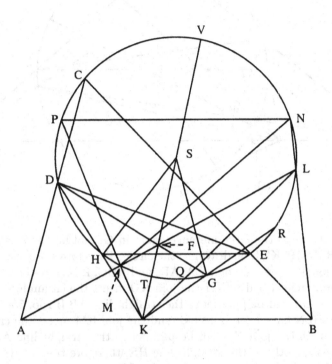

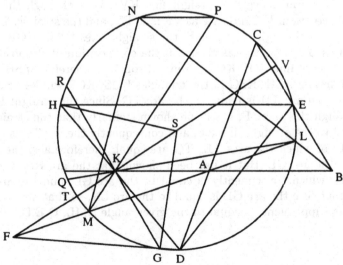

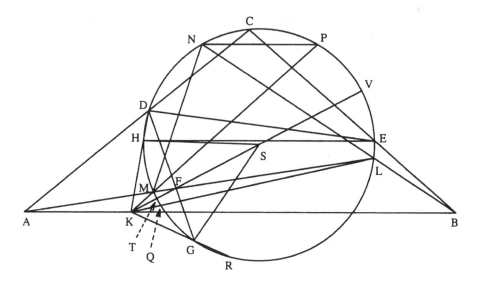

For let DF meet the circumference again in G, and let the straight lines KHD, KGR, KMP, KQL be drawn to the circle. Therefore since the rectangle BA, AK is equal to the rectangle CA, AD, and ADC, BEC have been inflected to the circumference, and in fact the join KD meets the circumference again in H, the join EH will be parallel to the straight line AB [Prop. 55]. Likewise since AML, BLN have been inflected, and the join KM meets the circumference again in P, the join NP will be parallel to the straight line AB [Prop. 55], and consequently to the straight line HE; therefore the arc HP is equal to the arc EN. And since the straight line KV passes through the centre, and, by construction, VK is to KT as VF is to FT, and the straight line DFG has been drawn through the point F, the straight lines KHD, KGR will cut off equal arcs HD, GR, as was shown in the composition of Proposition 53; hence the straight lines KH, KG are equal [Prop. 52]. Therefore if SH, SG are drawn, the angles KSH, KSG in the triangles KHS, KGS will be equal [I 8], so that also the arc TH is equal to the arc TG. Since the straight line ML passes through the point F, it will be shown similarly that the straight line KP is equal to the straight line KL, and consequently the arc TP is equal to the arc TL, of which the arcs TH, TG are equal; therefore also the arc HP is equal to the arc GL. But it has been shown that the arc HP is equal to the arc EN, which consequently is equal to the arc GL, and the arc EL is common, therefore the arc GE is equal to the arc LN, so that also the angle EDG, or its supplement, is equal to the given angle NML. Q.E.D.

Proposition 58

(This is the second part of the preceding general Porism.)

"Here the straight line which joins the points of intersection of the inflected lines and the circumference is parallel to a straight line which is given in position."

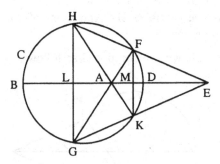

Let there be two points A, E, and from them straight lines AF, EF inflected in any manner to the circumference CHD which is given in position, and meeting it again in G, H; let the join GH be parallel to a straight line which is given in position. Let AH be joined and let it meet the circumference again in K. Therefore since AH, EH have been inflected from the points A, E to the point H on the circumference, and they meet it again in K, F, then, by hypothesis, the join KF will be parallel to the straight line which is given in position, so that it will also be parallel to GH. Again, EK having been joined, since AK, EK have been inflected from the points A, E to the point K, and AK meets the circumference again in H, the straight line which joins the point H and the other intersection of EK and the circumference will be parallel to FK by hypothesis; but the straight line HG is parallel to the straight line FK, therefore G is the point of intersection of EK and the circumference, that is the points E, K, G are in a straight line. Therefore since GH, KF are parallel, the triangles GFH, GKH will be equal (n) to each other [I 37], and, triangle GAH having been taken away from both, the triangle HAF will be equal to the triangle GAK. Now, AE having been joined, the triangle HAF is to the triangle HAE (as the straight line HF is to the straight line HE [VI 1], that is, on account of the parallels, as GK is to GE, that is [VI 1]) as the triangle GAK is to the triangle GAE; and the triangles HAF, GAK have been shown to be equal, therefore the triangles HAE, GAE are also equal. Let the straight line EA meet the circumference in D, B, and the straight lines HG, FK in the points L, M. Therefore the triangle EAH is to the triangle HAL and the triangle EAG is to the triangle GAL as the straight line EA is to the straight line AL; and the triangles EAH, EAG are equal, therefore the triangles HAL, GAL are equal; hence their bases HL, LG are also equal, and consequently the straight lines FM, MK are also equal. Therefore since the straight line LM bisects the two parallel lines GH, FK in the circle, LM will pass through the centre, and it will cut GH, FK at right angles. And since HG is perpendicular to the diameter BD of the circle, and GAF, HFE have been drawn, meeting the diameter in the points A, E, as BE is to ED so BA will be to AD [Prop. 51]. Therefore in order that the straight line which joins the points of intersection of the straight lines inflected from the points A, E and the circumference be parallel to a straight line which is given in position, the points A, E must be located on the diameter in such a way that BA is to AD as BE is to ED. And if one of the points A, E is given, the other will also be given.

It will be composed as follows.

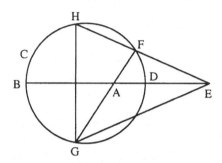

Therefore let the point A be given either inside or outside the circle BCD which is given in position, and through A let the diameter BAD be drawn on which let the point E be taken such that BE is to ED as BA is to AD; if straight lines AF, EF are inflected in any manner from the points A, E to the circumference and meet it again in G, H, the join GH will be perpendicular to the diameter BD.

For let EG be joined; and since BA is to AD as BE is to ED, and the straight line GAF has been drawn through the point A, the straight lines EFH, EG will cut off equal arcs, as has been shown in the composition of Proposition 53, so that by Corollary 2 of the same Proposition GH is perpendicular to the diameter BD.

Although this composition is contained in Proposition 53 and its Corollary 2, it seemed appropriate to present the investigation of this Proposition as well as that of the remaining parts of the general Porism.

(p.194)   **Proposition 59**

"Suppose that from two points A, B outside a circle straight lines AC, BC are inflected to the circumference, meeting it again in D, E, and that the rectangles CA, AD and CB, BE are together equal to the square of the join AB; let the perpendicular MK be drawn from the centre M to AB; then the rectangle BA, AK will be equal to the rectangle CA, AD. And conversely, if the rectangles CA, AD and CB, BE are together equal to the square of AB, and the point K is taken on AB making the rectangle BA, AK equal to the rectangle CA, AD, the straight line KM drawn to the centre will be perpendicular to AB."

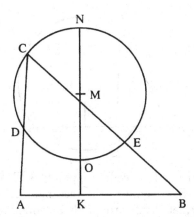

Let KM meet the circumference in the points N, O; therefore the rectangle CA, AD is equal to the square of AK and the rectangle OK, KN. Likewise the rectangle EB, BC is equal to the square of BK and the rectangle OK, KN. Therefore the rectangle CA, AD along with the rectangle EB, BC, that is, by hypothesis, the square of AB, is equal to the squares of AK, KB and twice the rectangle OK, KN. Let the squares of AK, KB be taken away, and twice the rectangle AK, KB will be equal to twice the rectangle OK, KN [II 4]; hence the rectangle AK, KB is equal to the rectangle OK, KN (n). Let the square of AK be added to both, and the rectangle BA, AK will be equal to the rectangle OK, KN and the square of AK, that is to the rectangle CA, AD. It will be shown similarly that the rectangle AB, BK is equal to the rectangle CB, BE.

And if the rectangles CA, AD and CB, BE are together equal to the square of AB, and the point K is taken on AB making the rectangle BA, AK equal to the rectangle CA, AD, the straight line KM drawn to the centre will be perpendicular to AB.

For if this is not the case, let the straight line MR perpendicular to AB be drawn from the centre; therefore, by the above, the rectangle BA, AR is equal to the rectangle CA, AD, that is to the rectangle BA, AK, which is impossible. Therefore MK is perpendicular to AB.

Proposition 60

"Suppose that from two points A, B, of which A is outside while B is inside the circle, straight lines AC, BC are inflected to the circumference, meeting it again in D, E, and that the rectangle CA, AD is equal to the rectangle CB, BE along with the square of the join AB; and from the centre M let MK be drawn perpendicular to AB; the rectangle BA, AK will be equal to the rectangle CA, AD. And conversely, if the rectangle CA, AD is equal to the rectangle CB, BE along with the square of AB, and the point K is taken on

AB making the rectangle BA, AK equal to the rectangle CA, AD, then the straight line KM drawn to the centre will be at right angles to AB."

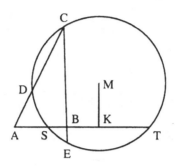

Let AB meet the circumference in the points S, T; and since the rectangle CA, AD, that is TA, AS, is equal to the rectangle CB, BE, or SB, BT, along with the square of AB, let the square of SK be added to both sides; therefore the square of AK will be equal to the rectangle SB, BT along with the squares of AB, SK [II 6]. Let the square of AB be taken away from both sides, and twice the rectangle AB, BK along with the square of BK will be equal to the rectangle SB, BT along with the square of SK [II 4]. Again let the square of BK be taken away from both sides, and twice the rectangle AB, BK will be equal to twice the rectangle SB, BT [II 5]; therefore the rectangle AB, BK is equal to the rectangle SB, BT. Let the square of AB be added to both sides, and the rectangle BA, AK will be equal to the rectangle SB, BT, or CB, BE, along with the square of AB [II 3], that is, by hypothesis, to the rectangle CA, AD.

The second part of the Proposition will be shown in the same way as the second part of the preceding Proposition.

Now Propositions 59, 60 are Lemmas for the following.

(p.195) # Proposition 61

(This is the third part of the general Porism.)

"Here the straight line which joins the points of intersection of the inflected lines and the circumference goes to a given point."

Case 1. Here the straight line which passes through the two given points does not meet the circle.

Let A, B be the two given points, and when two straight lines AC, BC have been inflected in any manner from them to the circumference CDE which is given in position, meeting it again in D, E, the join DE will go to a given point by hypothesis. Let that point be F, and let BD be joined, which meets the circumference again in G; and since the straight lines AD, BD have been inflected from the points A, B to D and meet the circumference again in C, G, the join CG will go to the same point F by hypothesis. Again, AG having been drawn, since the straight lines AG, BG have been inflected from the points A, B to G and BG meets the circumference again in D, the straight line which joins the point D and the other point of intersection of the straight line AG and the circumference will pass through the point F by hypothesis; therefore the straight line DF passes through that point of intersection; but the straight line DF meets the circumference in E, so that the point E is the point of intersection of AG and the circumference, that is the points A, G, E are in a straight line.

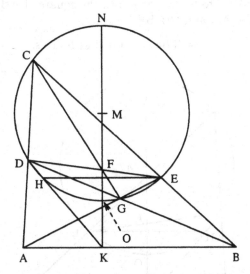

Let EH be drawn parallel to the straight line AB, and let it meet the circumference again in H, and let the join DH meet the straight line AB in K. Therefore since AC, BC have been inflected from the points A, B to the circumference, meeting it again in D, E, and EH has been drawn parallel to AB, and the join DH meets AB in K, the rectangle BA, AK will be equal to the rectangle CA, AD [Prop. 55]; therefore the points B, K, D, C are on a circle, and consequently the angle AKD is equal to the angle (DCE, that is to the angle) AGD, because the points E, G, D, C are also in a circle; therefore the points A, K, G, D are also in a circle, and consequently the rectangle AB, BK is equal to the rectangle DB, BG, that is the rectangle CB, BE.

Therefore since it has been shown that the rectangle BA, AK is equal to the rectangle CA, AD, and the rectangle AB, BK is equal to the rectangle CB, BE, the rectangles CA, AD and CB, BE together will be equal to the rectangles BA, AK and AB, BK, that is the square of AB [II 2]. Let M be the centre of the circle, and let MK be joined, which meets the circumference in the points N, O; therefore MK is perpendicular to AB [Prop. 59], and consequently to HE. Therefore since HE is perpendicular to the diameter NO, and HD has been drawn to the circumference, and meets the diameter in K, the straight line ED will pass through a given point on the diameter, namely the point which divides it in the ratio which NK has to KO, as is established by Proposition 51; but, by hypothesis, DE passes through the given point F; therefore F is on the diameter; for there cannot be two given points on the straight line DE, for if there were, the straight line DE would be given in position, which cannot be the case, since AC, BC have been drawn in any manner to the circumference. Therefore DE passes through the given point. Moreover the point F will be found by making NF to FO as NK to KO [Prop. 51]. Therefore in order that the straight line DE which joins the points of intersection of the lines inflected from the points A, B and the circumference should go to a given point, the square of AB must be equal to the rectangles CA, AD and CB, BE together.

It will be composed as follows.

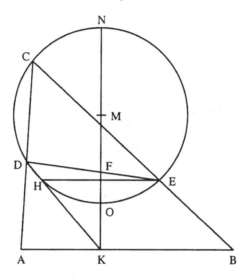

Let two points A, B be given and suppose that when straight lines AC, BC are inflected in any manner to the circumference CDE and meet it again in D, E, the square of the join AB is equal to the rectangle CA, AD and the rectangle CB, BE together; the join DE will go to a given point F, which is found by drawing through the centre M the straight line MK perpendicular to AB and meeting the circumference in N, O, and making NF to FO as NK to KO.

For let KD be joined and let it meet the circumference again in H; let DE meet the diameter in F, and let HE be joined. Therefore since the rectangle CA, AD along with the rectangle CB, BE is equal to the square of AB, and MK has been drawn perpendicular to AB, the rectangle BA, AK will be equal to the rectangle CA, AD [Prop. 59]; therefore HE is parallel to the straight line AB [Prop. 55], and consequently it is perpendicular to the diameter NMO. And since HD meets the diameter in K, and DE meets it in F, then NF will be to FO as NK is to KO [Prop. 51]; but NK, KO are given, therefore the point F to which DE goes is given. Q.E.D.

Corollary. And if the point A is given, and the straight line AB is given in position, the point B will be found by drawing MK perpendicular to AB, and taking the point B which makes the rectangle AK, KB equal to the rectangle OK, KN, as is clear from Proposition 59 (n1).

Case 2. Here the straight line which passes through the given points meets the circle.

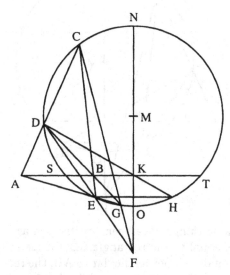

It will be shown, as in Case 1 and with the same words, that the rectangle BA, AK is equal to the rectangle CA, AD, and the rectangle AB, BK is equal to the rectangle CB, BE. Therefore since the rectangle CA, AD is equal (to the rectangle BA, AK, that is to the rectangle AB, BK along with the square of AB, that is) to the rectangle CB, BE along with the square of AB, if MK is drawn from the centre M, it will be perpendicular to AB [Prop. 60], and consequently to HE. Therefore since HE is perpendicular to the diameter NO, and HD has been drawn to the circumference, and meets the diameter in K, the straight line ED will pass through a given point on the diameter produced, as is established by Proposition 51; but, by hypothesis, DE passes through the given point F; therefore F is on the diameter produced; for there cannot be two given points on the straight line DE, since if there were, DE would be given in position, which is not the case. Therefore DE goes through the given point F, namely the point which makes NK to KO as NF to FO. Therefore in order that the straight line DE which joins the points of intersection of the inflected lines AC, BC and the circumference should go to a given point, the rectangle CA, AD must be equal to the rectangle CB, BE along with the square of AB.

It will be composed as follows.

Let two points A, B be given, and suppose that when straight lines AC, BC are inflected in any manner to the circumference CDE which is given in position and meet it again in D, E, the rectangle CA, AD is equal to the rectangle CB, BE along with the square of the join AB; the join DE will go to the given point F, which is found as in Case 1.

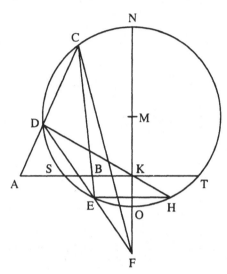

For, when the same things have been constructed as in Case 1, since the rectangle CA, AD is equal to the rectangle CB, BE along with the square of AB, and MK has been drawn perpendicular to AB, the rectangle BA, AK will be equal to the rectangle CA, AD [Prop. 60]; let DK be joined and let it meet the circumference again in H, and let DE meet the diameter in F; then the join HE will be parallel to the straight line AB [Prop. 55], and consequently it is at right angles to the diameter NMO. And since HD meets the diameter in K, and DE meets it in F, as NK is to KO so NF will be to FO [Prop. 51]; but NK, KO are given, therefore the point F to which the straight line DE goes is given.

Corollary. And since it has been shown in Proposition 60 (n1) that the rectangle AB, BK is equal to the rectangle SB, BT, when the square of BK has been added to both, the rectangle AK, KB will be equal to the square of SK which is given. Therefore if one of the points A, B, say A, is given, and the straight line AB is given in position, the other point B will be obtained by making the rectangle AK, KB equal to the square of SK.

Corollary 2 (to both cases). And if the straight line MK is drawn from the centre of the circle M perpendicular to the straight line AB, on which there are the two points A, B making the rectangle BA, AK equal to the rectangle CA, AD, any straight line ADC having been drawn to the circle, and BC is joined, which meets the circumference again in E, and DE is joined, which meets the diameter NMO which passes through K in F, then NK will be to KO as NF is to FO. Or, what is the same, if NK is to KO as NF is to FO, then DE will meet the diameter NO in F.

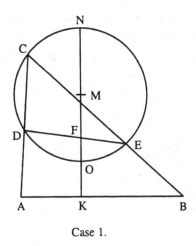

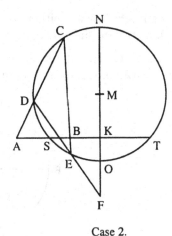

Case 1.   Case 2.

Case 1. For since the rectangle BA, AK is equal to the rectangle CA, AD, that is the square of AK along with the rectangle OK, KN, if the square of AK is taken away from both, and the square of BK is added to both, the rectangle AB, BK will be equal to the square of BK along with the rectangle OK, KN, that is to the rectangle CB, BE; and when the equal quantities have been added (n2), the square of AB will be equal to the rectangle CA, AD along with the rectangle CB, BE. Therefore, by the composition of this Case, NF will be to FO as NK is to KO.

Case 2. Here the straight line AB meets the circle. Since the rectangle BA, AK is equal to the rectangle CA, AD, that is the rectangle SA, AT, if the square of SK is added to both, the rectangle BA, AK along with the square of SK will be equal to the square of AK [II 6]; let the rectangle BA, AK be taken away from both sides, and the square of SK will be equal to the rectangle AK, KB [II 2]; and if the square of BK is taken away, the remaining rectangle AB, BK will be equal to the remaining rectangle SB, BT [II 35], that is the rectangle CB, BE, and if the square of AB is added, the rectangle BA, AK, that is the rectangle CA, AD, will be equal to the rectangle CB, BE along with the square of AB. Therefore, by the composition of Case 2, NF will be to FO as NK is to KO.

195)   **Proposition 62**

(This is a case of the general Porism, namely
the final one of Book 3 of Euclid's Porisms.)

"Suppose that a straight line has been drawn in any manner from a given point to a circle which is given in position; then from one or other of the points in which it meets the circumference, let a straight line be drawn parallel to a straight line which is given in position, and let the straight line be drawn from the point in which the first line drawn meets the circumference again to the point in which the other line drawn meets the same circumference again;

this straight line will either go to a given point or it will contain a given angle with a straight line which goes to a given point."

Part 1. Here the straight line joining the points of intersection of the lines drawn and the circumference goes to a given point.

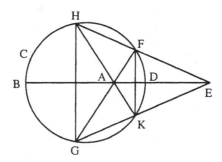

Let the point E be given, and let the straight line EFH be drawn in any manner from it to the circle BCD which is given in position, and from one or other of the points in which it meets the circumference, say from the point H, let the straight line HG be drawn parallel to a straight line which is given in position, and let it meet the circumference again in G, and let FG be joined; by hypothesis, FG will go to a given point; let this point be A, and from the point F let FK be drawn parallel to the straight line which is given in position, and consequently parallel to the straight line HG, and let it meet the circumference again in K. Therefore since the straight line EFH has been drawn from the point E to the circumference, and from the point F the straight line FK has been drawn parallel to the straight line which is given in position, the join HK will pass through the point A by hypothesis. Again, since EK has been drawn from the point E to the circumference, and KF is parallel to the straight line which is given in position, the straight line which joins the point F and the other point of intersection of the straight line EK and the circumference will pass through the point A by hypothesis; but the straight line FA passes through the point G on the circumference; therefore the point G is the point of intersection of the straight line EK and the circumference, that is the points E, K, G are in a straight line. Therefore since HG, FK are parallel, by what has been shown in Proposition 58 (n1), the straight line EA will bisect at right angles the straight lines HG, FK, and will pass through the centre of the circle. Let EA meet the circumference in the points B, D, and since EFH has been drawn to the circle from the point E, and HG is perpendicular to the diameter, and the join GF meets the diameter in A, as BE is to ED so BA will be to AD [Prop. 51]. But BE, ED are given, so that the ratio BA to AD is given; and BD is given, therefore the point A is given. Moreover it is clear that the straight line which is given in position and to which HG is parallel is at right angles to the diameter which passes through E.

It will be composed as follows.

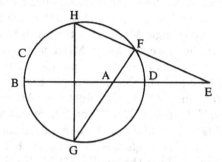

Let the straight line EDB be drawn through the centre from the point E, and let BA be made to AD as BE is to ED; then A will be the point which has to be found. For if EFH is drawn in any manner from the point E to the circle, and HG is drawn perpendicular to the diameter BD, the join GF will pass through the point A by Proposition 51.

Part 2. Here the straight line joining the points of intersection of the lines drawn and the circumference will contain a given angle with a straight line which goes to a given point.

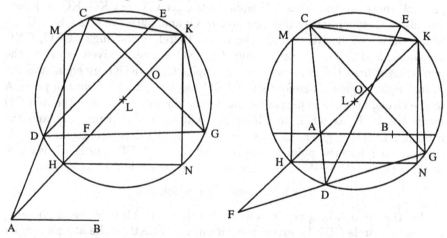

Let the point A be given, and let the straight line AB be given in position; then let the straight line ADC be drawn in any manner from the point A to the circle CDE which is given in position, and from one or other of the points in which it meets the circumference, say from the point C, let the straight line CE be drawn parallel to the straight line AB and let it meet the circle again in E, and let DE be joined; therefore, by hypothesis, DE will contain a given angle with a straight line which goes to a given point; let that point be F, and let DF be joined and let it meet the circumference again in G; therefore the angle EDG is given. Let AF be joined and let it meet the circumference in the points H, K, and let KM, HN be drawn parallel to AB, or to the straight line CE, and let them meet the circumference again in M, N; and let HM, KN be joined. Therefore since AK has been drawn from the point A to the circumference, and KM is parallel to AB which is given in position, MH will contain the given angle MHK with the straight line HK which goes to the given point F; therefore the angle MHK is equal to the given angle EDG. Likewise, since HN has been drawn parallel to the straight line AB, the angle NKH will be equal (to the same angle EDG, that is) to the angle MHK. But in the case in which the straight line AB meets the circle (see second figure above), if the point D, or H, or K is on the smaller segment cut off by the straight line AB, in place of the angle EDF, or MHF, or NKF the corresponding supplementary angle has to be taken, FD or FH having been produced (n2). Therefore since the angles MHK, NKH are equal in every case, MH, NK are parallel; and MK, HN are parallel, so that MHNK is a parallelogram, and consequently its opposite angles HMK, KNH are equal to each other; and MHNK is in a circle, therefore the angles HMK, KNH are equal to two right angles [III 22]; therefore both of them are right angles, and consequently HK passes through the centre of the circle [IV 5]. Let the centre be L, therefore the straight line AL is given in position. Then since the angle EDG is equal to the angle MHK, the arc EKG will be equal to the arc MEK, of which the arc EK is equal to the arc MC, because MK, CE are parallel, therefore the arc KG is equal to the arc CK. Let KG, KC be joined and also CG which meets the diameter in O; and since the arcs CK, KG are equal, the arcs CH, HG will also be equal; therefore the angles CKO, GKO are equal, and the two straight lines CK, KO are respectively equal to the two straight lines GK, KO, so that the angles COK, KOG are equal and are in fact right angles. Therefore since ADC has been drawn from the point A to the circle, and CG is perpendicular to the diameter HK, and the join GD meets HK in F, as KF is to FH so KA will be to AH [Prop. 51]; and the ratio KA to AH is given, so that the ratio KF to FH is also given; and HK is given, therefore the point F is given; and the angle EDF, or EDG, is equal to the given angle MHK. These are the things which had to be shown.

It will be composed as follows.

Let the point A be given, and let the straight line AB be given in position, and let the circle CDE be given in position, but let AB not be at right angles to the diameter which passes through the point A; through the centre L let AHK be drawn, and let KF be made to FH as KA is to AH, and let KM be drawn parallel to the straight line AB, and let HM be joined. If ADC is drawn in any manner from the point A to the circle, and CE is drawn parallel to AB, and DE, DF are joined, DF meeting the circumference again in G, the angle EDF will be equal to the given angle MHF.

For since KA is to AH as KF is to FH, and AD, FD have been inflected from the points A, F to the circumference, meeting it again in C, G, the join CG will be perpendicular to the diameter HK [Cor. 2 of Prop. 53]. Therefore the joins CK, KG are equal, so that the arcs CK, KG are also equal; and since CE, MK are parallel, the arcs MC, EK are equal; therefore also the arc MCK is equal to the arc EKG, and consequently the angle standing upon this, viz. EDG, is equal to the angle MHK, or the angle EDF is equal to the angle MHF.

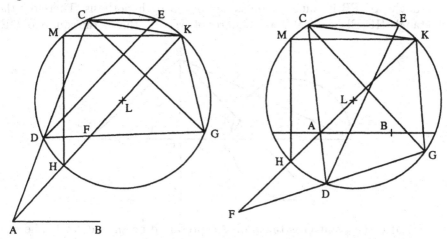

Proposition 63

(This is Proposition 159 of Pappus's Book 7.)

"Let there be a circle about the diameter AB, and let AB be produced, and let it be perpendicular to any straight line DE; moreover let the square of FG be put equal to the rectangle AF, FB. I say that if any point is taken 'on DE', as E, and the straight line drawn from it to the point G is produced, 'and meets the circumference in the points K, H', then also the rectangle HE, EK is equal to the square of EG."

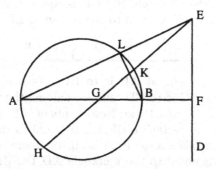

Let AE be joined "and let it meet the circumference again in L", and let BL be joined; therefore the angle ALB will be a right angle; but the angle AFE is also a right angle, "therefore the points B, L, E, F are on a circle, so that the rectangle EA, AL is equal to the rectangle FA, AB. But the rectangle EA, AL along with the rectangle AE, EL is equal (to the square of AE [II 2],

that is) to the squares of AF, FE [I 47], of which quantities EA, AL is equal
to the rectangle FA, AB"; therefore the remaining rectangle AE, EL is equal
to the remaining part, namely the rectangle AF, FB [II 2] and the square
of FE. Now the rectangle AE, EL is equal to the rectangle HE, EK, and the
rectangle AF, FB is equal to the square of FG by hypothesis. Therefore the
rectangle HE, EK is equal to the squares of GF, FE, that is the square of GE.

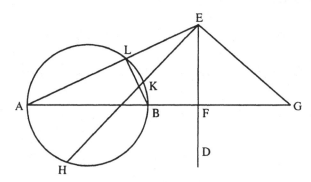

"And if the point G is taken on AF produced on the side of F, the rest
remaining as before, and any point E is taken on the straight line DE, from
which the straight line EKH be drawn in any manner to the circle, and EG
be joined, then the rectangle HE, EK will be equal to the square of EG." This
is demonstrated with the same words as the preceding part.

(p.196) ## Proposition 64

(This is Proposition 160 of Pappus's Book 7.)

"Let AD be to DC as AB is to BC, and let AC be bisected in the point E.
I say that three things hold, namely: the rectangle BE, ED is equal to the
square of EC; then the rectangle BD, DE is equal to the rectangle AD, DC;
and the rectangle AB, BC is equal to the rectangle EB, BD."

A   E   D   C   B

For since AD is to DC as AB is to BC, *componendo*, AB along with
BC will be to BC as AC is to CD; and by halving the preceding ratios, EB
will be to BC as EC is to CD; and *convertendo*, CE will be to ED as BE
is to EC; therefore the rectangle BE, ED is equal to the square of EC. Let
the square of ED be taken away from both, therefore the rectangle BD, DE
which is left [II 3] is equal to the rectangle AD, DC [II 5]. Again since the
rectangle BE, ED is equal to the square of EC, let both be taken away from
the square of EB; therefore the remaining rectangle EB, BD [II 2] is equal to
the rectangle AB, BC [II 6].

But now let the rectangle BD, DE be equal to the rectangle AD, DC; and
let AC be bisected in E. I say that AD is to DC as AB is to BC.

For since the rectangle BD, DE is equal to the rectangle AD, DC, let the square of DE be added to both, and the whole rectangle BE, ED [II 3] will be equal to the square of EC [II 5]. Therefore BE is to EC as EC is to ED. Therefore *convertendo*, and by doubling the preceding ratios, and *dividendo*, AD is to DC as AB is to BC.

.196)   ## Proposition 65

(This is Proposition 161 of Pappus's Book 7.)

These things being so, "let there be a circle about the diameter AB, and let AB be produced so that it is perpendicular to any straight line DE at the point F; and let AG be made to GB as AF is to FB. I say again that, if any point is taken 'on DE', as E, and the join EG meets the circumference in H, K, then HE is to EK as HG is to GK."

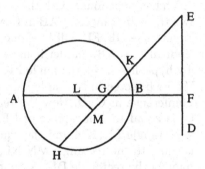

Let L, the centre of the circle, be taken, and from it let LM be drawn perpendicular to EH; then KM will be equal to MH [III 3]. Now since the two angles at M, F are right angles, the points E, F, L, M will be on a circle; "therefore the rectangle FG, GL is equal to the rectangle EG, GM". But the rectangle FG, GL is equal to the rectangle AG, GB, since AG is to GB as AF is to FB, and AB is bisected in the point L [Prop. 64]. Therefore the rectangle EG, GM is equal to the rectangle AG, GB, that is to the rectangle HG, GK, because AB, HK cut each other in the circle [III 35]; and HK is bisected in M, therefore from what has just been demonstrated [Prop. 64], as HE is to EK so HG will be to GK.

The three immediately preceding Propositions are of use for the two following, which are Porisms.

.196)   ## Proposition 66

"Suppose that the straight line AB and the circle CD have been given in position; the point will be given with the property that, if a straight line meeting the straight line AB which is given in position and the circle is drawn through it in any manner, the rectangle contained by the segments of the drawn line between the straight line AB and the points in which it meets the circumference will be equal to the square of the segment of the same drawn line between the straight line AB and that point which has to be shown given."

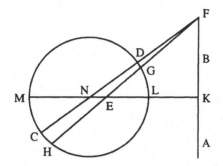

Let E be the point which has to be found with the property that, whenever a straight line EF is drawn through it meeting AB in F, and the circumference in the points G, H, the rectangle GF, FH will be equal to the square of EF. Let EK be drawn perpendicular to AB, and let it meet the circumference in L, M; again therefore, by hypothesis, the rectangle LK, KM will be equal to the square of EK. Let ML be bisected in N, and let the join FN be produced, and let it meet the circumference in C, D. Now the rectangle GF, FH, that is the rectangle DF, FC, is equal to (the square of EF, that is the squares of EK, KF, that is) the rectangle LK, KM and the square of KF; moreover the rectangle CN, ND is equal to the rectangle MN, NL, or the square of NL [III 35]; thus the sum, namely the rectangle DF, FC along with the rectangle CN, ND, will be equal to (the sum, namely the rectangle LK, KM and the squares of KF, NL, that is (n) the squares of NK, KF [II 6], that is) the square of NF. Let the rectangle NF, FD be taken away from both sides, and the remaining part, namely the rectangle CN, DF along with the rectangle CN, ND, that is the rectangle CN, NF, will be equal to the remaining rectangle DN, NF. Therefore the straight line CN is equal to the straight line ND; and since the straight lines ML, CD in the circle bisect each other in the point N, then N will be the centre of the circle [III 4], which is certainly given. But the straight line AB is given in position, and the angle NKB is a right angle, therefore NK is given in position, so that the points K, L, M are given. Therefore the rectangle LK, KM is given, that is the square of EK is given, and so KE will be given in position and magnitude; and the point K is given, therefore the point E is also given. This is what had to be shown.

It will be composed as follows.

Let N, the centre of the circle, be taken, and from it let NK be drawn perpendicular to AB and meeting the circumference in L, M; then let the mean proportional KE between MK and KL be found, which is located from the point K towards N; then E will be the point which has to be found.

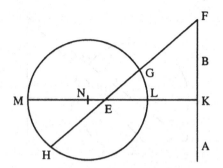

For if any straight line EF is drawn from it to the straight line AB and meets the circumference in G, H, the rectangle GF, FH will be equal to the square of EF, which has in fact been shown in Proposition 63.

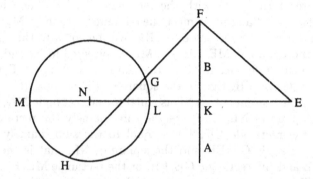

And if the straight line KE is located on the side opposite to that on which it was located in the preceding composition, and the straight line EF is drawn in any manner from the point E to the straight line AB, and then the straight line FGH is drawn in any manner from the point F to the circle, the rectangle GF, FH will be equal to the square of EF; for the straight line EF in this case is equal to the straight line EF in the previous case.

(196)

Proposition 67

"Suppose that the straight line AB and the circle CD have been given in position; the point will be given with the property that, if a straight line which meets the straight line AB and the circle is drawn through it in any manner, the rectangle contained by the segments of the drawn line between the straight line AB and the circumference will be equal to the square of the segment of the same drawn line between the straight line AB and the point which has to be shown given, along with the rectangle contained by the segments of the same drawn line between that point and the circumference."

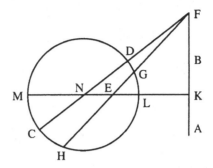

Let E be the point which has to be found with the property that, when a straight line EF is drawn through it in any manner meeting AB in F, and the circumference in the points G, H, the rectangle GF, FH will be equal to the square of EF along with the rectangle GE, EH. Let EK be drawn perpendicular to AB and let it meet the circumference in L, M; again therefore, by hypothesis, the rectangle LK, KM will be equal to the square of EK along with the rectangle ME, EL. Let ML be bisected in N, and let the join FN meet the circumference in C, D; therefore the rectangle CF, FD is equal to the rectangle GF, FH, that is the square of EF along with the rectangle GE, EH; and the rectangle CN, ND is equal to the rectangle MN, NL [III 35], that is the square of NL; therefore the sum, namely the rectangle CF, FD along with the rectangle CN, ND is equal to the sum, namely the square of EF, the rectangle GE, EH and the square of NL, that is the squares of FK, KE, NL and the rectangle GE, EH, or the rectangle ME, EL, that is the squares of FK, NL and the rectangle LK, KM, which is equal to the square of EK and the rectangle ME, EL. Therefore the rectangle CF, FD along with the rectangle CN, ND is equal to (the squares of FK, NL, and the rectangle MK, KL, that is the squares of FK, KN [II 6], that is) the square of NF. Let the rectangle NF, FD be taken away from both sides, and the remaining part, namely the rectangle CN, DF along with the rectangle CN, ND, that is the rectangle CN, NF, will be equal to the remaining rectangle DN, NF; therefore the straight line CN is equal to the straight line ND. And since the straight lines ML, CD in the circle bisect each other in the point N, then N will be the centre of the circle [III 4], which is certainly given. Therefore the straight line NK is given in position. Now since the rectangle LK, KM is equal to the square of EK along with the rectangle ME, EL, let the square of NL be added to both sides, and the sum, namely (n1) the square of NK, will be equal to (the sum, namely the squares of EK, NL and the rectangle ME, EL, that is) (n2) the squares of EK, NE and twice the rectangle ME, EL [II 5]. Let the square of EK along with the square of NE be taken away from both sides, and the remaining part, namely (n3) twice the rectangle NE, EK [II 4] will be equal to twice the rectangle ME, EL; hence the rectangle NE, EK is equal to the rectangle ME, EL. Let the square of NE be added to both, and the rectangle KN, NE will be equal to the square of NL, which is given. But KN is given, hence the point E is given.

It will be composed as follows.

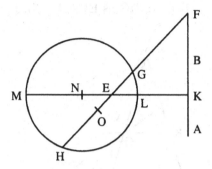

From the centre N let NK be drawn perpendicular to AB, and let it meet the circumference in L, M; and let the third proportional NE be found for KN, NL; then E will be the point which has to be found.

For let the straight line EF be drawn to AB in any manner, and let it meet the circumference in G, H; the rectangle GF, FH will be equal to the square EF along with the rectangle GE, EH. For the rectangle KN, NE is equal to the square of NL, and, the square of NE having been taken away from both, the rectangle NE, EK is equal to the rectangle ME, EL, so that, by Proposition 64 at the end of its demonstration, as MK is to KL so ME will be to EL; therefore, by Proposition 65, as HF is to FG so HE is to EG; and if HG is bisected in O, the rectangle OE, EF will be equal to the rectangle HE, EG by Proposition 64, and, by the same Proposition, the rectangle HF, FG will be equal to the rectangle OF, FE, that is the square of EF and the rectangle OE, EF, that is the square of EF and the rectangle HE, EG. Q.E.D.

And these are the Loci and Porisms, namely those which are contained among the preceding Propositions from the seventh up to this one, which I was able to determine to be Euclid's from Pappus's quite imperfect and mutilated description, and by means of the Lemmas which were composed for these things by that most expert Geometer, to whom almost all that is now known about the Analysis of the Ancients is due; for Pappus alone preserved the names and arguments of the books which the Ancients wrote about it. However the description which he gives of the Porisms is so brief and obscure, and spoiled by the injury of time or other causes, that unless God had given generously spirit and strength to enquire perseveringly into it, they would have lain hid from Geometers perhaps for ever.

Now in order that everything which Pappus preserved concerning this matter may be presented together, we shall add below the rest of his Lemmas for the Porisms, with a view to pleasing especially those who desire to look further into these things.

(p.197)   # Proposition 68

(This is Pappus's Proposition 145 of Book 7.)

"From some point E let straight lines EF, EB be drawn to three straight lines AB, AC, AD; and let EH be to HG as EF is to FG; I say that ED is to DC as EB is to BC."

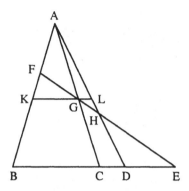

Through G let KL be drawn parallel to BE; now EH is to HG as EF is to FG; moreover EB is to GK as EF is to FG, and DE is to GL as EH is to HG; thus DE will be to GL as EB is to GK; and *permutando*, KG will be to GL as BE is to ED. But BC is to CD as KG is to GL; therefore BC is to CD as BE is to ED; and *permutando*, ED is to DC as EB is to BC. And the things which relate to the cases will be explained similarly.

Serenus of Antissa demonstrates this Proposition in Proposition 33 of *On the Section of a Cylinder*, but much less elegantly.

Corollary. The following locus is easily deduced from this Proposition of Pappus, viz.

If a straight line EHF is drawn from a given point E to two straight lines AB, AD which are given in position, and it is cut in G so that EH is to HG as EF is to FG, then the point G will lie on a straight line which is given in position.

For let the point B be taken on one of the straight lines which are given in position; let EB be joined, and let the join AG meet it in C and let AD meet it in D. Therefore since EF is to FG as EH is to HG, then by this Proposition, EB will be to BC as ED is to DC; and *permutando*, BC will be to CD as BE is to ED. But the ratio BE to ED is given, because the straight lines themselves are given, and so the ratio BC to CD is given; and BD is given, therefore the point C is given. And the point A is given, therefore the straight line AC on which the point G lies is given in position.

Composition.

Let EB be drawn to the given point B on AB, and let it meet AD in D; and let BC be made to CD as BE is to ED; the join AC will be the straight line on which the point G lies. For let the straight line EHF be drawn in any manner, and let it meet AC in G. Since BE is to ED as BC is to CD, and *permutando*, ED is to DC as EB is to BC, then by this Proposition, EH will be to HG as EF is to FG.

And the converse of the Locus is the following. If straight lines EDB, EHF are drawn to AB, AD, and BD, FH are cut in C, G such that EB is to BC as ED is to DC and EF is to FG as EH is to HG, the points C, G, A will be in a straight line. For let KL be drawn through G parallel to BE; therefore EF is to FG as EB is to KG; and EH is to HG as DE is to GL; thus DE is to GL as EB is to KG, and *permutando*, KG is to GL as BE is to ED; but BC is to CD as BE is to ED, because EB is to BC as ED is to DC. Therefore BC is to CD as KG is to GL. Therefore the points C, G, A are in a straight line by the Lemma at note [b] in Proposition 23 of this work.

Proposition 69

(This is Proposition 146 of Pappus's Book 7.)

"Let there be two triangles ABC, DEF which have equal angles at A, D. I say that the triangle ABC is to the triangle DEF as the rectangle BA, AC is to the rectangle ED, DF."

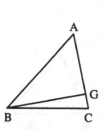 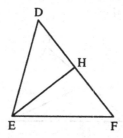

For let perpendiculars BG, EH be drawn "to AC, DF", and since the angle at A is equal to the angle at D, and the angles at G, H are right angles, DE will be to EH as AB is to BG. But the rectangle BA, AC is to the rectangle BG, AC as AB is to BG. And the rectangle ED, DF is to the rectangle EH, DF as DE is to EH. Therefore the rectangle ED, DF is to the rectangle EH, DF as the rectangle BA, AC is to the rectangle BG, AC, and *permutando* (the rectangle BG, AC is to the rectangle EH, DF as the rectangle BA, AC is to the rectangle ED, DF). But as the rectangle BG, AC is to the rectangle EH, DF, so the triangle ABC is to the triangle DEF, "for each rectangle is twice the corresponding triangle"; and therefore the triangle ABC is to the triangle EDF as the rectangle BA, AC is to the rectangle ED, DF.

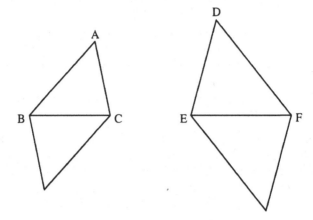

It may be shown more briefly thus. Let the parallelograms with sides AB, AC and DE, DF be completed; these, and consequently half of them, the triangles ABC, DEF, have the ratio compounded from the ratios of the sides by [VI 23], and the rectangles BA, AC and ED, DF have the same ratio; therefore as rectangle is to rectangle, so triangle is to triangle.

Proposition 70

(This is Proposition 147 of Pappus's Book 7.)

"Let the angles at A and D be equal to two right angles. I say again that the triangle ABC is to the triangle DEF as the rectangle BA, AC is to the rectangle ED, DF."

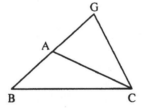 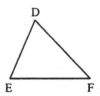

Let BA be produced, and let AG be put equal to BA, and let GC be joined. And so since the angles at A and D are equal to two right angles, and in fact the angles BAC, CAG are likewise equal to two right angles, the angle CAG will be equal to the angle at D. Therefore the triangle ACG is to the triangle DEF as the rectangle GA, AC is to the rectangle ED, DF. But GA is equal to AB, and the triangle AGC is equal to the triangle ABC. Therefore the triangle ABC is to the triangle EDF as the rectangle BA, AC is to the rectangle ED, DF.

Proposition 71

197)

(This is Proposition 152 of Pappus's Book 7.)

"Let the square of AD be to the square of DC as AB is to BC. I say that the rectangle AB, BC is equal to the square of BD."

A   C   B   D   E

For let DE be put equal to CD; therefore (n1) the rectangle EA, AC along with the square of CD, that is along with the rectangle CD, DE, is equal to the square of AD [II 6]. And so since the square of AD is to the square of DC as AB is to BC, *dividendo*, the rectangle EA, AC will be to the rectangle CD, DE as AC is to CB, that is as the rectangle EA, AC is to the rectangle EA, BC. Therefore the rectangle EA, BC is equal to the rectangle CD, DE. Therefore, on account of proportion* and *dividendo*, DB is to BC as AD is to DE, that is to DC, and so the remaining part AB is to the remaining part BD as DB is to BC (n2). Therefore the rectangle AB, BC will be equal to the square of BD.

197)

Proposition 72

(This is Proposition 153 of Pappus's Book 7.)

"Again let the square of AD be to the square of DC as AB is to BC. I say that the rectangle AB, BC is equal to the square of BD."

A   E   D   C   B

Likewise of course let DE be put equal to DC; the rectangle CA, AE along with the square of CD, that is along with the rectangle ED, DC will be equal to the square of AD [II 6]. And, *dividendo*, the rectangle EA, AC will be to the rectangle ED, DC as AC is to CB, that is as the rectangle EA, AC is to the rectangle EA, CB. Therefore the rectangle AE, CB is equal to the rectangle ED, DC; and, on account of proportion† and *componendo*, DB is to BC as AD is to DE, or DC; hence the sum AB is to the sum DB as DB is to BC. Therefore the rectangle AB, BC is equal to the square of DB.

Proposition 73

(This is Proposition 154 of Pappus's Book 7.)

"Let the straight lines AD, DC be tangent to the circle ABC, and let AC be joined; 'let the straight line DB be drawn to the circle in any manner and let it meet the circumference in B, E, and the straight line AC in F'. I say that BF is to FE as BD is to DE."

---

* AE is to ED as DC is to CB.

† AE is to ED as DC is to CB.

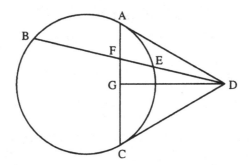

For since AD is equal to DC, the rectangle AF, FC along with the square of FD will be equal to the square of DA. * But the rectangle AF, FC is equal to the rectangle BF, FE; and the square of DA is equal to the rectangle BD, DE. Therefore the rectangle BF, FE along with the square of FD is equal to the rectangle BD, DE. Since this is so, BF will be to FE as BD is to DE. †

(p.197)

Proposition 74

(This is Proposition 157 of Pappus's Book 7.)

"Let there be a semicircle on the straight line AB, and from the points A, B let the straight lines BD, AE be drawn at right angles to the straight line ACB; let DE be drawn in any manner and let it meet the circumference in F; moreover let FG be drawn from the point F at right angles to DE and let it meet AB in the point G. I say that the rectangle AE, BD is equal to the rectangle AG, GB."

"Analysis."

"Let it be so"; therefore GB will be to BD as EA is to AG, that is the sides about the equal angles are proportional; thus, "GE, GD having been joined", the angle AGE is equal to the angle BDG [VI 6]. But the angle AGE

---

* Let DG be drawn perpendicular to AC; therefore since AD is equal to DC, then AG will be equal to GC. Therefore the rectangle AF, FC along with the square of FG is equal to the square of AG [II 5]; let the square of GD be added to both sides, and the rectangle AF, FC along with the squares of FG, GD, that is along with the square of FD, will be equal to the squares of AG, GD, that is the square of DA.

† For since the rectangle BF, FE along with the square of FD is equal to the rectangle BD, DE, let these equal quantities be taken away from the rectangle BD, DF, and the remaining rectangle BF, DE will be equal to the remaining rectangle BD, FE. Therefore as BD is to DE so BF is to FE.

Commandino's figure and commentary put the straight line DB drawn through the centre. However it is clear from Pappus's demonstration that the straight line DB is to be drawn to the circle in any manner.

is indeed equal to the angle AFE, which is contained in the same segment, "for the points E, A, F, G are on a circle on account of the right angles EAG, EFG"; now the angle BDG is equal to the angle BFG, which is in the same segment, "for again the points B, G, D, F are on a circle on account of the right angles GBD, GFD". Therefore the angle AFE is equal to the angle BFG. This holds in fact, since both angles AFB, EFG are right angles.

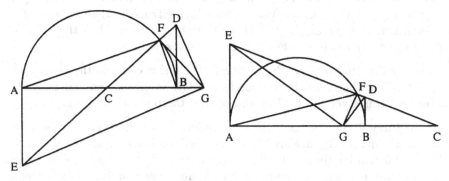

Composition.

Since both angles AFB, EFG are right angles, if the angle EFB is taken away from both, the angle AFE will be equal to the angle BFG. But the angle AFE is in fact equal to the angle AGE, which is contained in the same segment; moreover, for the same reason, the angle BDG is equal to the angle BFG. Therefore the angle AGE is equal to the angle BDG; and the right angle EAG is equal to the right angle GBD; hence also the remaining angle is equal to the remaining angle, and the two triangles will be similar. Therefore as EA is to AG so GB is to BD, and consequently the rectangle AE, BD is equal to the rectangle AG, GB.

The three Corollaries below follow from this Proposition.

Corollary 1. Suppose that from the given point G the straight line GBA is drawn perpendicular to two parallel straight lines AE, BD which are given in position, and the straight line DE is drawn between these two lines cutting off segments AE, BD which contain a rectangle equal to the given rectangle AG, GB; moreover let GF be drawn from the point G perpendicular to DE; the point F will lie on a circumference which is given in position.

For let AF, DG, GE, FB be joined; and since the rectangle AE, BD is equal to the rectangle AG, GB, then GB will be to BD as EA is to AG, and they are about equal angles which are right angles, therefore the triangles EAG, GBD are equiangular, so that the angle AGE is equal to the angle BDG. And since the angles EAG, EFG are right angles, the points E, A, F, G are on a circle; therefore the angle AFE is equal to the angle AGE in the same segment. Similarly since the angles GBD, GFD are right angles, the angle BFG will be equal to the angle BDG. Therefore the angle AFE is equal to the angle BFG, and consequently the angle AFB is equal to the angle

EFG, namely a right angle. Therefore since the straight line AB is given in postion and magnitude, the point F will lie on the circumference described on the diameter AB.

Corollary 2. And if the straight lines AE, BD touch the circle AFB at the extremities of the diameter AB which is given in position and magnitude, and EFD is drawn cutting off segments AE, BD which contain a rectangle equal to a given area, and from the point F in which ED meets the circumference FG is drawn at right angles to ED, and meets AB in G, then the point G will be given, that is to say FG will go to a given point.

For, by this Proposition, the rectangle AG, GB is equal to the given rectangle AE, BD; therefore the rectangle AG, GB is given, and AB is given, therefore AG is given [*Dat.* 85 or 86], so that the point G is also given.

Corollary 3. Suppose that the straight line AB has been given in position and magnitude; if AE, BD are drawn tangent to the circle on the diameter AB, a point will be given with the property that, if a straight line is drawn from it in any manner to a point on the circumference, and a straight line is drawn from this point at right angles to the previously drawn line, the last line drawn will cut off from the tangents segments adjacent to the extremities of the diameter which will contain a rectangle equal to a given rectangle.

Now the point which has to be found will either be on the diameter AB or off it. First suppose that it is off the diameter; therefore if a straight line is drawn from it to one or other of the extremities of the diameter, and from this extremity a straight line is drawn at right angles to this line, it will cut off no segment from the tangent which is drawn through this extremity, but it will cut off a segment from the other tangent, hence there will be no rectangle contained by the segments, which is contrary to hypothesis; thus the point which has to be found is not off the diameter.

Therefore let the point which has to be found be on the diameter AB, and let it be G, and when GF has been drawn in any manner to the circumference, let the straight line EFD be drawn through the point F at right angles to the straight line GF and let it meet the tangents in E, D; therefore, by hypothesis, the rectangle contained by AE, BD will be equal to the given rectangle; but, by this Proposition, the same rectangle is equal to the rectangle AG, GB, which consequently will be given; and AB is given, so that the point G will be given [*Dat.* 85 or 86].

Therefore let the point G be taken on the diameter AB which will make the rectangle AG, GB equal to the given rectangle; then G will be the point which has to be found; for if GF, EFD have been drawn as described, the rectangle AE, BD will be equal to the rectangle AG, GB, that is, by construction, to the given rectangle.

Proposition 75

(This is Proposition 158 of Pappus's Book 7.)

"Let ABC be a triangle, having side AB equal to side AC,[†] and let AB be produced to D, and from the point D let DE be drawn making the triangle BDE equal to the triangle ABC. I say that if that one of the equal sides which is transverse to the equal triangle is bisected by the 'straight' line BF 'which meets the straight line DE in G', then the square of AF is to the square of FH as the sum of FB, BG is to FG."

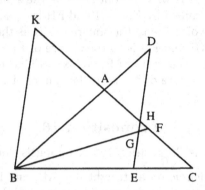

"Suppose that it is true, and" let BK be drawn through B parallel to DE, and let CA be produced to K; therefore the square of AF is to the square of FH as the sum of FK, KH is to FH, that is as the rectangle contained by the sum of FK, KH and FH is to the square of FH.[A] Therefore what is contained by the sum of FK, KH and FH, that is the excess of the squares of FK, KH, is equal to the square of AF.[B] Therefore the excess of the squares of KF, FA is the square of KH. But the excess of the squares of KF, FA is the rectangle CK, KA [II 6], "for AC is bisected in F"; therefore the rectangle CK, KA is equal to the square of KH; then consequently as CK is to KH, that is as CB is to BE, so HK is to KA, that is as DB is to BA. This is in fact the case, "for since the triangles DBE, ABC are equal, and have the angle DBC in common, DB is to BA as CB is to BE" [VI 15].[C]

[†] There is nothing in the demonstration which depends upon the equality of the sides AB, AC, for it holds in any triangle; therefore without doubt the Proposition is corrupted.

[A] For BK, GH are parallel, therefore the sum of FK, KH is to FH as the sum of FB, BG is to FG, that is, by hypothesis, as the square of AF is to that of FH.

[B] For if HK is produced so that the produced part is equal to HK itself, the matter will be clear by [II 6].

[C] The material contained within inverted commas is briefer than Pappus's version which it replaces. The same material is changed in the composition.

Now it will be composed in the following manner.

Since the triangle DBE is equal to the triangle ABC, and they are about the common angle DBC, as CB is to BE, that is as CK is to KH, so DB will be to BA, that is so HK will be to KA [VI 15]; therefore HK is to KA as CK is to KH. Hence the rectangle CK, KA is equal to the square of KH. But the rectangle CK, KA is the excess of the squares of KF, FA, therefore also the excess of the squares of KF, FA is the square of KH, and because of this the square of FA is the excess of the squares of FK, KH. But the excess of the squares of FK, KH is what is contained by the sum of FK, KH and FH; therefore what is contained by FK, KH and FH is equal to the square of FA. Therefore the square of AF is to the square of FH as the rectangle contained by the sum of FK, KH and FH is to the square of FH, that is as the sum of FK, KH is to FH. But the sum of FB, BG is to FG as the sum of FK, KH is to FH; therefore the square of AF is to the square of FH as the sum of FB, BG is to FG. Q.E.D.

Proposition 76

(This is Proposition 162 of Pappus's Book 7.)

"Let there be a semicircle on a straight line AB, and let CD be parallel to AB, and let CE, DG be drawn perpendicular 'to AB'. I say that AE is equal to GB."

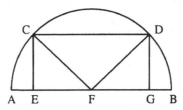

Let F, the centre of the circle, be taken, and let CF, FD be joined. Therefore CF is equal to FD, and consequently the square of CF will be equal to the square of FD. But in fact the squares of CE, EF are equal to the square of CF; moreover the squares of DG, GF are equal to the square of FD. Therefore the squares of CE, EF are equal to the squares of DG, GF; since the square of CE is equal to the square of DG, the remaining square of EF is therefore equal to the remaining square of FG, and so the straight line EF is equal to the straight line FG; but also the whole of AF is equal to the whole of FB. Therefore also the remaining part AE is equal to the remaining part GB. Q.E.D.

It is appropriate to note the considerable difference between the demonstrations of this Proposition and the preceding Proposition 73. For the preceding demonstration seems to be exceedingly brief, and certain things had to be added to it; however this one is detailed exactly in the manner of the

Elements. But the same thing is to be observed in many other of Pappus's demonstrations.

Proposition 77

(This is Proposition 163 of Pappus's Book 7.)

"Let there be a semicircle on a straight line AB, and from any point C 'on it produced' let CD be drawn, and let the perpendicular DE be drawn. I say that the square of AC exceeds the square of CD by the quantity which is contained by both AC, CB and AE."

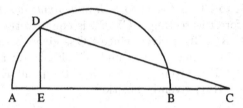

"Let it be so"; therefore the square of AC is equal to the square of CD, that is the squares of DE, EC, and the quantity which is contained by both AC, CB and AE. Hence, when the rectangle CA, AE has been taken away from both sides, the remaining rectangle AC, CE is equal to the square of DE, that is the rectangle AE, EB, and the square of CE, and what is contained by AE, CB. And again, when the square of CE has been taken away from both sides, the rectangle AE, EC which is left is equal to the rectangle AE, EB, and the quantity which is contained by AE, BC; this is indeed the case [II 1].
The composition, as Commandino observes, is clear.

And if the point C is taken between the points A, B, it will be shown in a not dissimilar way that the square of AC exceeds the square of CD by the quantity which is contained by the excess of AC, CB and AE, as long as AC is greater than CB. But if CB is greater than AC, the square of CD will exceed the square of AC by the quantity which is contained by the excess of CB, AC and AE.

198)
Proposition 78

(This is required for the Problem in the following Proposition.)

"Suppose that two straight lines AC, CD have been given in position, and the point E is given in the angle which is supplementary to the angle ACD; from the point E the straight line EGF has to be drawn which cuts off from the straight lines CA, CD segments CF, CG, adjacent to the point C, which contain a rectangle equal to a given area." *

---

* This is Case 1 of Locus 3 in Book 1 of Apollonius's *The Cutting off of an Area* as restored by the most distinguished Halley.

Suppose that it has been done; and from the point E let EH, EK be drawn parallel to AC, CD; and since the point E is given, EH, EK will be given. Therefore the rectangle HE, EK has a given ratio to the rectangle FC, CG, for both of them are given. Therefore as the side HC of the first is to the side GC of the second, so the remaining side CF of the second will be to the straight line to which the remaining side CK of the first has the given ratio, namely that which the rectangle HE, EK has to the rectangle FC, CG [*Dat.* 63]. Let this straight line be CL; therefore FC is to CL as HC is to CG (n1), and *convertendo*, CF is to FL as CH is to HG; but the triangles EKF, GHE are equiangular, therefore KF is to EH, that is KC, as EK, that is CH, is to HG; hence CF is to FL as FK is to KC, and *dividendo*, CL is to LF as FC is to CK. Therefore the rectangle CF, FL is equal to the rectangle KC, CL; but the rectangle KC, CL is given, since KC is given, as also is CL to which it has a given ratio. Hence the rectangle CF, FL is given, and CL is given, therefore the point F is given [*Dat.* 85], and the point E is given; therefore the straight line EGF is given in position.

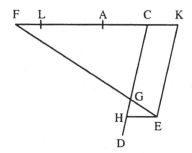

It will be composed as follows.

Let EH, EK be drawn parallel to AC, CD, and let the rectangle HC, CL be the given area to which the rectangle which is to be taken off by the straight line drawn from the point E is equal. Then let the rectangle CF, FL, equal to the rectangle KC, CL, be applied to the straight line CL, exceeding it by a square (n2), and let the straight line EF be drawn meeting the straight line CD in G; the rectangle FC, CG will be equal to the rectangle HC, CL, that is the given area. For since, by construction, the rectangle CF, FL is equal to the rectangle KC, CL, as FC is to CK so CL is to LF, and *componendo*, as FK is to KC so CF is to FL; but, because the triangles EKF, GHE are equiangular, FK is to EH, that is KC, as EK, that is CH, is to HG; therefore CF is to FL as CH is to HG, and *convertendo*, FC is to CL as HC is to CG; and consequently the rectangle FC, CG is equal to the rectangle HC, CL, that is the given area.

And it is clear that only the straight line EF resolves the Problem; for if the other point of application were taken which is on the straight line LC produced on the side of C, the straight line drawn from the point E to this point would indeed take off segments which contain a rectangle equal to the given area, not from the straight lines CA, CD however, but from those which are in the opposite directions.

Proposition 79

(This is Proposition 164 of Pappus's Book 7 and
his 38th and last Lemma for Euclid's Porisms.)

"The parallelogram AD having been given in position, to draw from a given point E a straight line EF, and to make the triangle FCG equal to the parallelogram."

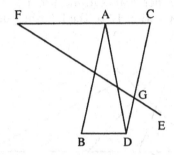

Now suppose that it has been done; the triangle FCG is then equal to the parallelogram AD, and the parallelogram AD is in fact twice the triangle ADC; thus the triangle FCG will also be twice the triangle ADC. But the rectangle FC, CG is to the rectangle AC, CD as the triangle is to the triangle, because they are located about the same angle at C [Prop. 69]. Moreover the rectangle AC, CD is given; therefore the rectangle FC, CG is also given. And the straight line EF has been drawn from the given point E to the straight lines AC, CD which are given in position, cutting off "straight lines FC, CG which contain" an area equal to the given area, namely twice the rectangle AC, CD. Therefore the straight line EF is "given" in position, by the preceding Proposition.

Moreover it will be composed as follows.

Let the parallelogram AD be given in position, and moreover let the point E be given; and from the point E let the straight line EF be drawn from the point E to FC, CD which are given in position, cutting off "straight lines FC, CG which contain" an area FC, CG equal to the given area, namely twice the rectangle AC, CD; and from the same things which were stated in the analysis, we will show that the triangle FCG is equal to the parallelogram AD. Therefore the straight line EF resolves the Problem. Moreover it is established that it alone does this, since it is also unique.

Notes on Part III

Note on Proposition 27 (p. 100). Compare this demonstration with Simson's alternative demonstration of his Proposition 17 (Pappus's Proposition 129) which is given here in the note on Proposition 17, p. 92. Simson noted that Proposition 142 is the converse of Proposition 129 and that the alternative demonstration is the converse of Pappus's demonstration of Proposition 142.

Note on Proposition 28 (p. 101). Simson notes that this Proposition is a case of Proposition 17 in his *Treatise*. Strictly it is a limiting case, since a certain point in his earlier diagram and proof has to become a point at infinity. If we apply Proposition 17 to the following diagram,

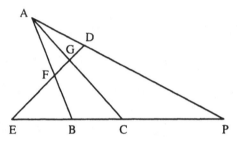

we obtain $\frac{EF.DG}{ED.GF} = \frac{EB.CP}{EP.BC}$. Now let $P \to \infty$. Then $\frac{CP}{EP} \to 1$, we obtain the diagram for the enunciation of Proposition 28 and we deduce that $\frac{EF.DG}{ED.GF} = \frac{EB}{BC}$, from which the required result follows.

In the enunciation the phrase in single quotation marks is Simson's addition to Commandino's text.

Note on Propositions 29 and 30 (pp. 102, 103). These two propositions constitute what is generally known as "Pappus's Theorem". MacLaurin noted that it and the Hyptios Porism (Proposition 19) are in fact special cases of his five-point construction of a conic (see [21, p. 153]).

Note on Proposition 31 (p. 104). Note that E lies on CD, although this is not specified in the enunciation. (n) From $\frac{CB}{BF} = \frac{CK}{GF}$ we deduce that $\frac{KC}{GF} = \frac{CB+CK}{BF+GF} = \frac{KB}{BG}$ as claimed.

Note on Proposition 34 (p. 106). The quotation from Pappus before the enunciation identifies Proposition 34 as a reconstruction of the sixth Porism which Pappus recorded, in extremely abbreviated form, from Book I (see p. 36). The enunciation is followed by a restatement in terms of named lines and points. Note that G and K are respectively the points of intersection of CE with AF and DE with BF. If the other possibility, namely CE with BF and DE with AF were considered, then M would have to be replaced by the point of intersection of AC and BD. In terms of the diagrams the Proposition asserts that as E varies on AB the line GK always passes through the point of intersection of the fixed lines AD, BC. Simson's argument should begin with the customary statement: "Suppose that the Porism is true." An alternative reconstruction of this sixth Porism by Simson is described in Appendix 2.

Note on Proposition 36 (p. 108). Proposition 35 is actually applied twice: first as described in the demonstration and secondly as it is enunciated.

Note on Proposition 37 (p. 109). The paragraph immediately preceding the diagram is a statement of what has to be demonstrated; the demonstration begins in the next paragraph.

In the subsequent application to give an alternative demonstration of Proposition 34 the lettering differs from that used in its previous demonstration – the lettering and diagram of Proposition 37 apply.

Note on Proposition 38 (p. 110). Simson's reference following the enunciation to "the main part of Pappus's Proposition on the four lines" presumably identifies Proposition 19 of the *Treatise*. (See also Appendix 1.)

Note on Proposition 39 (p. 112). Simson criticises Pappus and Commandino here, but there seems to be an error in his own version, which is probably typographical: Proposition 26 rather than Proposition 28 was cited in the proof of Proposition 39. In fact Commandino followed exactly the Greek text here (see Jones's *Pappus* [17, pp. 276–279, 465]).

Note on Proposition 40 (p. 113). (n1) In the application of Proposition 8, Case 3, the points A, B, C, D, E, H, K, L of Proposition 40 correspond to the points C, A, G, D, B, E, F, H respectively of Case 3, where it is shown that the locus of C is a straight line through F and $\frac{AG}{GH} = \frac{BG}{GA}$. Consequently we deduce that in Proposition 40 the locus of A is a straight line through K, which explains Simson's assertion "Let it be AK ...", and $\frac{BC}{CL} = \frac{EC}{CB}$ (i).
(n2) In the application of Proposition 8, Case 4, Article 2 (wrongly referred to as Article 3, which does not exist, in the original) the points A, B, C, D, E, F, G, K, L of Proposition 40 correspond to the points D, H, B, C, A, E, K, F, G respectively of Article 2, where it is shown that the locus of E is a straight line through F and $\frac{AB}{GH} = \frac{BK}{KG}$. For Proposition 40 this says that the locus of F is a straight line through K (Simson's "Let it be FK ...") and $\frac{EC}{LB} = \frac{CG}{GL}$ (ii). In the composition L and then G are determined from conditions (i) and (ii); note that since E, C, B are given, the ratio $\frac{EC}{CB}$ is given.

(n3) The converse of Proposition 39 asserts that $\frac{EB^2}{EC.CB} = \frac{BG}{GC}$, which is the identity quoted in Simson's reference to Euclid's having combined the two constructions into one. He then proceeds to deduce this identity from (i) and (ii). Simson cites [VI 1], which asserts that "triangles or parallelograms which have the same altitude, are to each other as their bases". Thus we have from $\frac{EB}{BL} = \frac{BC}{LC}$ (deduced from (i) *dividendo* and *permutando*) on comparing rectangles that $\frac{BC.CE}{EC.CL} = \frac{BC}{CL} = \frac{EB}{BL} = \frac{EB.EB}{EB.BL}$, so that $\frac{EB.BL}{EC.CL} = \frac{EB^2}{BC.CE}$ as claimed. The other application of [VI 1] is similar.

Note on Proposition 41 (p. 115). The diagram in Simson's text shows AB perpendicular to AC, but this is not required for the result or its demonstration. The Lemma to which Simson refers is contained in the footnote to Proposition 22 which is marked with † ‡ (p. 83).

(n) The equalities EG.FH = AG.AH = BG.CH give $\frac{BG}{FH} = \frac{EG}{CH}$, from which we deduce $\frac{BE}{FC} = \frac{BG-EG}{FH-CH} = \frac{BG}{FH}$, and $\frac{AH}{EG} = \frac{FH}{AG}$, from which we deduce $\frac{AF}{AE} = \frac{AH+FH}{EG+AG} = \frac{FH}{AG}$, as stated.

Note on Proposition 43 (p. 117). (n) From $\frac{AB}{BD} = \frac{DB}{BC}$, which is (*), we deduce $\frac{AB+DB}{BD+BC} = \frac{DB}{BC}$, that is $\frac{AD}{DC} = \frac{DB}{BC}$, which is (†).

Note on Proposition 45 (p. 118). (n) Note that since $\frac{DB}{BC} = \frac{AB}{BD}$ we can deduce that $\frac{AB-DB}{BD-BC}$ in Fig. 1 and $\frac{DB-AB}{BC-BD}$ in Fig. 2 are equal to $\frac{AB}{BD}$, so that in either case $\frac{AD}{DC} = \frac{AB}{BD}$, which is Simson's explanatory insertion at the beginning of the demonstration.

Note on Proposition 47 (p. 119). (n1) In the analysis of Case 1 we deduce from $DE^2 = CE^2 + AD^2 + F.EB$ and $DB^2 = CB^2 + AD^2$ that $DE^2 - DB^2 = CE^2 - CB^2 + F.EB$, hence $(DE + DB).(DE - DB) = (CE + CB).(CE - CB) + F.EB$, and by use of DB = GD, CB = HC this yields GE.BE = HE.BE + F.EB = (HE + F).BE. Thus GE = HE + F and so GH = F; since GH = GB - HB = 2(DB - CB) = 2DC, we have that F, which serves only to establish a fixed magnitude, is equal to 2DC as claimed.

(n2) In the application of the converse of Proposition 42 its points A, B, C, D are replaced respectively by the points B, A, C, D of Proposition 47. The composition is essentially a reversal of these steps.

(n3) In Case 1 of the Corollary, following Case 1 of the Proposition, we get from $DB^2 = CB^2 + AD^2$ that $AD^2 = DB^2 - CB^2 = (DB - CB).(DB + CB) = DC.(DC + 2CB) = DC^2 + 2DC.CB = DC^2 + 2DC.(CE + EB)$, from which it follows that $AD^2 + CE^2 = (DC + CE)^2 + 2DC.EB = DE^2 + 2DC.EB$ as claimed.

(n4) For Case 2 of this Corollary note that the identity $DC^2 + 2DC.CB + CE^2 = CE^2 + AD^2$, which is cited from Case 1, does not depend on the position of E. We now apply the fact that for both Fig. 1 and Fig. 2 we have $DE^2 + 2DC.CE = (DC - EC)^2 + 2DC.CE = DC^2 + EC^2$.

(n5) According to Simson, if CE > CB, Case 3 of the Proposition is dealt with in the same way as Case 1 except for details of the composition. As in Case 1 we have $2DB.EB + BE^2 = GE.EB = HE.EB + F.EB$. Now we use $HE.EB = HE.(HE + BH) = HE.(HE + 2CH) = HE^2 + 2HE.CH$ to get $2DB.EB + BE^2 = HE^2 + 2HE.CH + F.EB$. Finally we combine this with $DB^2 = BC^2 + AD^2 = CH^2 + AD^2$ to deduce that $DE^2 = CE^2 + AD^2 + F.EB$.

(n6) In the case $CE < CB$ Simson gives a separate argument (analysis and composition) which uses ideas similar to those already discussed. For the analysis we require $BE.EH + EC^2 = (BC + CE).(BC - CE) + EC^2 = BC^2$ and for the composition $BC^2 - HE.EB = BC^2 - (BC - CE).(BC + CE) = CE^2$. Note that if $CE < CB$ we can have E between B and C; although this situation is not depicted in the figures, the arguments given do hold for it.

Note on Proposition 49 (p. 130). In his alternative solution of the Problem Simson refers to the "new edition" of Euclid's *Data*, by which he means his own edition, which he appended to the second edition (1762) of the *Elements* [31]. In this instance the edition is important, because Simson rearranged some of the material and included several additional items, so that numbering does not always correspond to that of the original. He cites Proposition 92, which is Proposition 89 in the original and has the following enunciation: "If a straight line given in magnitude be drawn within a circle given in magnitude; it shall cut off a segment containing a given angle."

Note on Proposition 50 (p. 133). Simson's investigations of the last three Porisms of Euclid's Book 3 go back to the early stages of his work on Porisms. Volume F (pp. 48–49) of his *Adversaria* [43] contains an article dated 1 March 1723 in which he gives Pappus's Greek versions and then explains "the meaning of these porisms which we have discovered with much effort". However in later marginal notes he refers to articles dated 25 August 1727 and 14 May 1729 for their true meanings. The first of these is contained in Volume K (pp. 6–9) and in it Simson gives two versions of the "antepenultimate Porism" and also notes as a Corollary that the following "Apollonian locus" may be deduced: "Suppose that from a given point F two straight lines FD, FE are drawn which contain a given angle and have a given ratio to each other; moreover let the extremity of one D lie on a circle LG which is given in position; then the extremity of the other E will lie on a circle MH which is given in position." It also contains a reference to p. 88 of Volume I, where the Porism is presented in the form given in Proposition 50 – this article is not dated, but neighbouring dated articles suggest that it belongs to the period 1730–31. For an English version of Pappus's Greek statement which Simson quotes see p. 37. Concerning the article of 14 May 1729 see the note on Proposition 53.

As usual Simson's arguments are quite clear, but it may help to note the following points.

(n1) The term *convex* (resp. *concave*) is used to denote the portion of the circumference which is nearer to (resp. further from) the exterior point under consideration. More precisely, take the set of lines through the point which meet the circle in two distinct points; the convex (resp. concave) part is obtained by taking for each such line the intersection nearer to (resp. further from) the point. Although it is not specified in the enunciation, it is assumed in the analysis and composition that the points D, E are on the same portions of the circumferences relative to the point C.

(n2) (Analysis) Simson asserts that the angle ACB is given; this is because angle ACK (= angle HCK) has been constructed to be equal to the given angle and C, K, B have been shown to be collinear.

(n3) (Composition) Simson notes concerning triangles ADC, BEC that (a) angles ACD, BCE are equal, (b) $\frac{AC}{AD} = \frac{CB}{BE}$ and (c) angles ADC, BEC are *both* either greater than 90°, which is the case in the diagram, D and E being on the convex parts of the two circumferences, or less than 90°, which would happen if D and E were taken on the concave parts. These three conditions guarantee that the triangles are similar/equiangular.

(n4) (Corollary) The term "given in magnitude" applied to a triangle means that its area is given. Note that since the area of triangle CRE is $\frac{1}{2}$ RC.CE sin($\angle$RCE), the ratio of the rectangle RC.CE to the area of triangle CRE is $2 \operatorname{cosec}(\angle$RCE). Having shown that $\frac{EC}{CG} = \frac{DC}{CF}$, one should follow the analysis of the Proposition from this point. In three places I have added statements enclosed within angled brackets which seem to be necessary for the argument.

William Trail (1746–1831), to whom Simson attributes a version of the Corollary, is probably best remembered for his much criticised biography of Simson, *Account of the Life and Writings of Robert Simson, M.D.* [40]. He was a student at Marischal College, Aberdeen during 1759–63, after which he moved to Glasgow, where he graduated M.A. in 1766. In the same year he was appointed Professor of Mathematics at Marischal College. In 1770 he published *Elements of Algebra for the use of Students in Universities*, which became a popular textbook. However, in 1779 he resigned his Chair having taken orders in the Church of Ireland, in which he held several appointments, notably Chancellor of Connor.

Finally we should note that, according to the date which Simson added after the Corollary, he recorded this material when he was in his eightieth year.

Note on Proposition 51 (p. 137). This result asserts that {AB, HG} is a harmonic set.

Note on Proposition 52 (p. 138). Although Simson's diagram shows the point E outside the circle, the Proposition and the proof given continue to hold if E is inside the circle. The Proposition is applied in both situations in Proposition 54.

Note on Proposition 53 (p. 139). (See the first paragraph of the note on Proposition 50.) I have been unable to locate an article dated 14 May

1729. However on pp. 126–128 (reading in reverse order) of Volume L of the *Adversaria* [43] there is an article dated 15 May 1729 in which Simson presents Proposition 53 with slightly different but equivalent wording. It is curious that in this article he refers to the Proposition as the "antepenultimate Porism", since the accompanying Greek statement is clearly that of the "penultimate Porism". For an English version of Pappus's statement which Simson quotes see p. 37.

. (n) In the demonstration of Corollary 2 Simson refers to the composition of the Proposition, where he shows that "the angles at the centre BNH, BNG will be equal"; clearly angles BNH and LNH are equal, likewise angles BNG and LNG, so that angles LNH, LNG are equal as claimed.

Note on Proposition 54 (p. 142). For this Proposition Simson reproduces the first figure for Case 2 of Proposition 57 and also refers the reader to its second and third figures. It seems preferable however to give the Proposition its own diagrams free from unnecessary additional lines.

Note on Proposition 57 (p. 146). (Case 1) The four diagrams provided for the analysis show the four significant possibilities for the given points A, B: one inside, one outside the circle; both inside the circle; both outside the circle and on the same side of the centre; both outside the circle and on opposite sides of the centre. For the composition I have included a corresponding set of diagrams from which the lines and points which are only referred to in the analysis have been omitted. In several places in both the analysis and the composition Simson refers to alternative angles, e.g. (n3), "And since the angle BQE, or in certain cases TQE, has been shown to be equal (to the angle ACB, or DCE, that is) to the angle DGE in the same segment, ...". In such cases we have to refer to the individual diagrams. In the instance quoted, for the first three diagrams $\angle BQE = \angle TQE$ and $\angle ACB = \angle DCE$, whereas in the fourth diagram $\angle TQE$ is the supplement of $\angle BQE$ and $\angle DCE$ is the supplement of $\angle ACB$. In another instance we have (n5): "therefore the angle BEA, or in certain cases CES, is equal to the angle TVF, ..."; in the first and second diagrams $\angle BEA$ and $\angle CES$ are the same angle, in the third diagram they are opposite angles and therefore equal, while in the fourth diagram $\angle BEA$ and $\angle CES$ are supplementary angles with $\angle CES = \angle TVF$. Otherwise Simson's arguments appear to be quite clear, but it may help to note the following points.

(n1) That $\operatorname{arc} DM = \operatorname{arc} MK$ follows from $\operatorname{arc} DC = \operatorname{arc} KH$ and $\operatorname{arc} MC = \operatorname{arc} MH$.

(n2) The triangles BQE, ACB are equiangular since the angles at B are equal and from AB.BQ = CB.BE we obtain $\frac{AB}{BC} = \frac{EB}{BQ}$, i.e., the sides about these angles are in proportion.

(n4) Similarly we deduce that triangles QTC, FTV are equiangular from QT.TF = CT.TV, and then $\angle CQT = \angle TVF$ since they are corresponding angles in these triangles.

(Case 2) Simson shows three situations in his diagrams; apparently he did not consider it necessary to provide a diagram for the case where both A and B lie inside the circle – I have provided an appropriate diagram for the analysis of this case at the end of this note. As in Case 1, I have included a corresponding set of diagrams for the composition, although in this case there is little change – the lines OL, OB are not required, while SG, SH now appear. (n6) Again we have to check in each diagram to see whether we want an angle or its supplement when alternatives are mentioned. Simson calls attention to the angles EDF, NMF, OLF; in the first and third figures for both the analysis and the composition all three angles are equal, whereas in the second diagrams the angles EDF, NMF are equal to the supplement of the angle OLF; in the diagram at the end of this note the angles EDF, OLF are equal to the supplement of the angle NMF. Note that (n7) AK is determined from BA.AK = CA.AD, which is independent of the particular line ADC through A meeting the circumference in C, D, and that (n8) F is characterised as the harmonic conjugate of K with respect to the points V, T where the line through K and the centre of the circle meets the circumference.

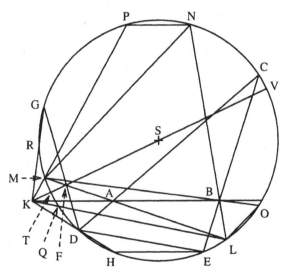

Note on Proposition 58 (p. 157). (n) Here the statement that two triangles are equal means that they have the same area.

Note on Proposition 59 (p. 158). (n) From $AB^2 = AK^2 + KB^2 + 2OK.KN$ and $AB^2 = (AK + KB)^2 = AK^2 + KB^2 + 2AK.KB$ we deduce that AK.KB = OK.KN. Then $AK^2 + AK.KB = AK.(AK + KB) = AK.AB$ and $AK^2 + OK.KN = CA.AD$. The last identity is a general result which Simson uses again in the second Corollary to Proposition 61 and in Propositions 90, 91 and 93. It may be established easily as follows (see diagram below).

We have CA.AD = NA.AJ and, since AKOJ is a cyclic quadrilateral (right angles at K, J), NA.NJ = NK.NO, that is NA.(NA − AJ) = NK.(NK − OK), so that CA.AD = $NA^2 − NK^2$ + NK.OK = AK^2 + OK.KN as required.

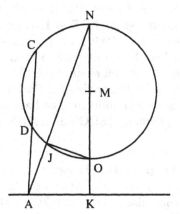

Note on Proposition 60 (p. 159). The argument is elementary, but it may be helpful to expand on the manipulations. From TA.AS = SB.BT + AB^2 we get SK^2 + TA.AS = SK^2 + SB.BT + AB^2. Now SK^2 + TA.AS = SK^2 + (AS + 2SK).AS = $SK^2 + AS^2$ + 2SK.AS = $(AS + SK)^2 = AK^2$, so that $AK^2 − AB^2 = SK^2$ + SB.BT. But $AK^2 − AB^2$ = (AK − AB).(AK + AB) = BK.(2AB + BK), which gives 2AB.BK + $BK^2 = SK^2$ + SB.BT. Then 2AB.BK = $SK^2 − BK^2$ + SB.BT = (SK − BK).(SK + BK) + SB.BT = SB.BT + SB.BT, i.e., AB.BK = SB.BT. Finally, AB^2 + AB.BK = AB.(AB + BK) = AB.AK and AB^2 + SB.BT = AB^2 + CB.BE = CA.AD by hypothesis.

Note on Proposition 61 (p. 160). (n1) In the Corollary at the end of Case 1 Simson refers to Proposition 59 in connection with AK.KB = OK.KN and in the first Corollary to Case 2 he refers to Proposition 60 for the identity AB.BK = SB.BT. These are actually found in the demonstrations of the cited Propositions. (See their notes above.)

Corollary 2 involves manipulations similar to those described in detail in the note on Proposition 60 and so it seems unnecessary to give a detailed account here. (n2) The statement in Case 1 of Corollary 2 "when the equal quantities have been added" means that the identities BA.AK = CA.AD and AB.BK = CB.BE, which have already been established, are to be added. This gives AB.(AK + BK) = AB^2 = CA.AD + CB.BE. Simson uses the result discussed in the note on Proposition 59 to express CA.AD as AK^2 + OK.KN.

Note on Proposition 62 (p. 165). Simson describes this Proposition as a "case of the general Porism" (*casus Porismatis generalis*). In view of the change of assumptions *variant* would seem to be a more appropriate term than *case*.

The situation considered in Part 1 may be regarded as a limiting case of that considered in Part 2. Note that in the composition of Part 2 Simson instructs, "but let AB not be at right angles to the diameter which passes through the point A". However, if we let $\angle BAK \to 90°$, then $E \to G$, since CG is at right angles to this diameter, and $DE \to DG$, which passes through the given point F; this corresponds to Part 1 with obvious changes in lettering.

(n1) In the analysis of Part 1 the reference to Proposition 58 is specifically to its analysis, beginning at "Therefore since GH, KF are parallel ...".

(n2) In the second figure for the analysis of Part 2 the points D, H lie on the smaller segment, so, as Simson indicates, we have to take the corresponding supplementary angles, namely $\angle EDG$ and $\angle MHK$ in place of $\angle EDF$ and $\angle MHF$ respectively.

Note on Proposition 63 (p. 169). The items contained within single quotation marks correspond to Simson's additions to Pappus's text. The statement "and let it be perpendicular to any straight line DE" in the enunciation means of course that DE is an arbitrary straight line perpendicular to the diameter AB produced. Note that the value of HE.EK does not depend on whether EKH passes through G.

Note on Proposition 64 (p. 170). Here $\{AC, BD\}$ is a harmonic set. For the first identity we use $AB + BC = AC + 2CB = 2(EC + CB) = 2EB$, for the second $EC + ED = AE + ED = AD$ and for the third $EB + EC = EB + AE = AB$. A converse is added after the proofs of the three identities; this is required for the composition of Proposition 67.

Note on Proposition 65 (p. 171). The statement "and let AB be produced so that it is perpendicular to any straight line DE at the point F" in the enunciation means that DE is an arbitrary straight line perpendicular to the diameter produced, F being their point of intersection. The Proposition asserts that $\{HK, GE\}$ is a harmonic set.

Note on Proposition 66 (p. 171). (n) In the analysis we use the fact that $LK.KM + NL^2 = (NK - NL).(NK + MN) + NL^2 = (NK - NL).(NK + NL) + NL^2 = NK^2$. Note that, just as in the corresponding situation in Proposition 63, the value of GF.FH does not depend on whether the straight line FGH passes through the given point E.

Note on Proposition 67 (p. 173). We need the following identities for the analysis:

(n1) $NL^2 + LK.KM = NL^2 + (NK - NL).(NK + MN) = NL^2 + (NK - NL).(NK + NL) = NK^2$;

(n2) $NL^2 = (NE + EL)^2 = NE^2 + 2NE.EL + EL^2 = NE^2 + (2NE + EL).EL = NE^2 + (NE + NL).EL = NE^2 + (MN + NE).EL = NE^2 + ME.EL$;

(n3) $NK^2 - EK^2 - NE^2 = (NK - EK).(NK + EK) - NE^2 = NE.(NK + EK) - NE^2 = NE.(NK + EK - NE) = 2NE.EK$.

In the composition E is characterised by the requirement $\frac{KN}{NL} = \frac{NL}{NE}$, i.e. $KN.NE = NL^2$; thus E is the pole of AB with respect to the given circle.

Note on Proposition 68 (p. 176). The demonstration relates to the configuration shown in the diagram, but of course there are other ways in which the lines may intersect. It is indicated that the demonstration will be similar in the other cases.

A Latin translation of Serenus's work was published by Commandino as *De Sectione Cylindri* in his *Apollonii Conicorum Libri Quarti* [4] and the Greek text was edited by Halley in his *Apollonii Pergaei Conicorum Libri Octo et Sereni Antissensis De Sectione Cylindri et Coni Libri Duo* [14]. Serenus's birthplace was given as Antissa by Halley, but it is now believed that he was born in Antinoupolis, Egypt. (See *Dictionary of Scientific Biography* XII, pp. 313–5 (1975).)

There is some vagueness in the demonstration of the Locus about what is given. In fact we just need to take any one transversal through E, keep it fixed, and relate the ratios for any other transversal to those on the chosen one. The Lemma to which Simson refers for the collinearity of C, G, A does not seem to be appropriate, although it does deal with a similar figure; however, since KL, BE are parallel it follows easily that AC divides KL in the same ratio as BC to CD, and so AC must pass through the point G, which has been shown to divide KL in this ratio.

Note on Propositions 71 and 72 (p. 179). These Propositions differ only in the relative positions of the points as shown by their respective diagrams. If A, B, C are given, the equation $\frac{AB}{BC} = \frac{AD^2}{DC^2}$ will in general determine two positions of the point D, which are considered in the two Propositions.

(n1) At the beginning of the demonstration of Proposition 71 we require $EA.AC + CD^2 = (AD + DE).(AD - CD) + CD^2 = (AD + CD).(AD - CD) + CD^2 = AD^2$. (n2) In its final step, from $\frac{AE}{ED} = \frac{DC}{CB}$ we obtain (*dividendo*) $\frac{AE-ED}{ED} = \frac{DC-CB}{CB}$, that is $\frac{AD}{DE} = \frac{DB}{BC}$, and then $\frac{AD-DB}{DE-BC} = \frac{DB}{BC}$, that is $\frac{AB}{BD} = \frac{DB}{BC}$, since DE = CD.

The demonstration of Proposition 72 follows the same pattern, except that in the final step *dividendo* is replaced by *componendo* and addition rather than subtraction is used to get $\frac{DB}{BC} = \frac{AD+DB}{DC+BC} = \frac{AB}{DB}$.

Note on Proposition 74 (p. 180). The Corollaries have been added by Simson.

Note on Proposition 75 (p. 183). In place of the instruction contained in Simson's footnote (*B*) we could simply note that $(FK + KH).FH = (FK + KH).(FK - KH) = FK^2 - KH^2$. We also need $KF^2 - FA^2 = (KF +$

FA).(KF − FA) = (KF + FC).(KF − FA) = CK.KA. In both the analysis and the composition Simson could have referred to Proposition 69 rather than [VI 15].

Note on Proposition 78 (p. 185). Halley's restoration of Apollonius's *The Cutting off of an Area* is contained in the same volume as his account of Pappus's Preface to Book 7, viz. his *Apollonii Pergaei de Sectione Rationis Libri Duo* [13].

(n1) In the analysis we construct CL such that $\frac{CK}{CL}$ is the same as the given ratio $\frac{HE.EK}{FC.CG}$, or equivalently $\frac{CK.HC}{FC.CG}$, so that CL = $\frac{FC.CG}{HC}$. (n2) In the composition we obtain CL as (given area)/HC and then we require a point F on CL such that CF.FL = KC.CL. According to his terminology, Simson invokes the technique of "application of areas" to find F ([VI 29]; see also [16, pp. 100–103]). In this case a rectangle of area KC.CL has to be constructed on CL produced in such a way that the part of the figure beyond L is a square.

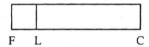

F   L   C

Note that from the fact that CF.FL and CL are given we actually get two positions for F (Fig. 1). As Simson notes at the end of the demonstration, if we take the other point F* then we still get the required rectangle but the segments are in the opposite directions (Fig. 2).

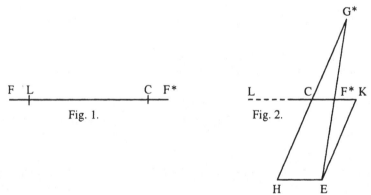

Fig. 1.   Fig. 2.

We have CF*.F*L = KC.CL, so that $\frac{CF^*}{KC} = \frac{CL}{F^*L}$ and hence $\frac{CF^*}{KC-CF^*} = \frac{CL}{F^*L-CL}$, that is $\frac{CF^*}{F^*K} = \frac{CL}{CF^*}$. But $\frac{CG^*}{EK} = \frac{CF^*}{F^*K}$ since the triangles CF*G*, F*EK are similar. Hence $\frac{CL}{CF^*} = \frac{CG^*}{EK}$ and therefore CG*.CF* = CL.EK = HC.CL as required.

Part IV. Various Porisms:
Fermat, Simson and Stewart

In Propositions 80–85 Simson discusses at length four of Fermat's Propositions. Some of this material originated with Matthew Stewart, as did the last four Propositions (90–93).

Among the Varia Opera Mathematica of the most skillful Geometer Pierre de Fermat, which were published at Toulouse in the year 1679 after his death, there are on pages 117, 118 five Propositions which he calls Porisms; the second of these is about a certain property of the Parabola, which we have demonstrated in Proposition 19 of the Fifth Book in the second edition of *Conic Sections*; but this and also the first, third and fifth, as Fermat enunciates them, are nothing but Theorems, while the fourth is a Datum. Indeed he says there is nothing to prevent them being transformed into Problems, and provides an example in his fifth Proposition, which he has turned into a Problem, certain things having been suppressed in the enunciation, but the Problem is in fact a Porism; and he adds in his fifth Proposition: "and the construction is not hard to see from the Theorem stated above (namely in his 5th Proposition), for if the ratio of the straight line AZ to the fourth part of ZD is set equal to a given ratio, all things will be determined." Whence it is clear that Fermat had not investigated the solution of this Porism at all, but had deduced its construction from the Theorem which he had come upon. And the same has to be said about the rest of his Propositions. But since, as far as I know, no one has in fact demonstrated these Propositions as Theorems, let alone as Porisms, we have investigated them in the following manner, where they have been expressed in the form of Porisms.

Proposition 80

(This is Fermat's first Proposition.)

"Suppose that two straight lines AB, AC have been given in position and that two points D, E have been given outside them; let straight lines DF, EF be inflected in any manner from these points to one of the straight lines AC, meeting the other straight line AB which is given in position in G, H; then the inflected straight lines will cut off from AB two segments adjacent to given points, which will contain a given rectangle; the given points and the rectangle have to be found."

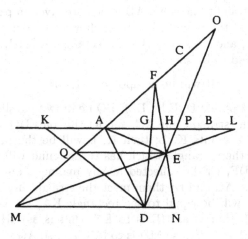

Suppose that the Porism is true, and that K, L are those points to which the segments cut off by the straight lines DF, EF are adjacent; now it has to be shown that the points K, L are given, and also that the rectangle contained by the segments cut off which are adjacent to these points is given. Therefore, by hypothesis, the rectangle KG, HL is given. Again let straight lines DA, AE be inflected from the same points D, E to the point A in which the straight lines which are given in position AB, AC meet each other; therefore the rectangle KA, AL will be given by hypothesis, and consequently the same rectangle is equal to the rectangle KG, HL. Therefore AL is to LH as GK is to KA, and *dividendo*, AH is to HL as GA is to AK; and *permutando*, KA is to HL as GA is to AH. Let DM be drawn to AC parallel to AB, and so the point M is given; further let FE meet MD in N; therefore GA is to AH and KA is to HL as DM is to MN. Let DE be joined and having been produced let it meet AC, AB in O, P; and since the straight lines DO, EO which meet the straight line AB in P have been inflected from the points D, E to AC, the rectangle KP, PL will be equal, by hypothesis, to the rectangle KA, AL. Therefore AL is to LP as PK is to KA, and *dividendo*, AP is to PL as PA is to AK; therefore KA, PL are equal. Now it was shown that DM is to MN as KA, that is PL, is to HL, consequently *permutando*, MN is to HL as MD is to PL, and [V 19] MD is to PL as (DN is to HP, that is as) DE is to EP. And the angles MDE, EPL are equal, therefore, ME, EL having been joined, the triangles MDE, EPL are equiangular, consequently the angle DEM is equal to the angle PEL; therefore the points M, E, L are in a straight line. But the points M, E are given, so that the straight line MEL and the point L are given in position. Let DK be joined and let it meet the straight line AM in Q; therefore KQ is to QD as (KA, or PL, is to MD, that is as) PE is to ED; consequently the join EQ is parallel to MD; and the point E is given, and the straight line MD is given in position; therefore EQ is given in position, and the point Q will be given, but the point D is also given; therefore the point K, in which the straight lines DQ, AB which are given in position meet each other, is given; and the point L has been shown to be given. Therefore the rectangle KA, AL, and consequently the rectangle KG, HL are given. These things are what had to be shown.

It will be composed as follows.

From the given points D, E let DM, EQ be drawn parallel to the straight line AB to meet the straight line AC, and let the joins DQ, ME be produced to meet AB in the points K, L; then K, L will be the points which have to be found, and the rectangle which has to be found will be the rectangle KA, AL. For let DF, EF be inflected in any manner from the points D, E to the straight line AC and let them meet the straight line AB in G, H; the rectangle KG, HL will be equal to the rectangle KA, AL. For KA is to MD as (AQ is to QM, that is as HE is to EN, that is as) HL is to MN; and *permutando*, as KA is to HL, so MD is to MN, and so AG is to AH; therefore (n1) KG is to AL as KA is to HL [V 12]. Consequently the rectangle KG, HL is equal to the rectangle KA, AL. Moreover it is to be noted that the straight line DE should not be parallel to the straight line AC (n2).

.243)

Proposition 81

(This is Fermat's third Proposition, but more generally stated.)

"Suppose that from a given point A on the circumference of a circle which is given in position any straight line AB is drawn, meeting in E the straight line CD which is given in position and is inscribed in the circle and meeting the circle again in B; the point with the following property will be given: if the straight line is drawn to it from the point B, meeting the straight line CD in F, then the ratio of the rectangle CF, ED to the rectangle CE, FD will be the same as the given ratio which the straight line GH has to the straight line GK."

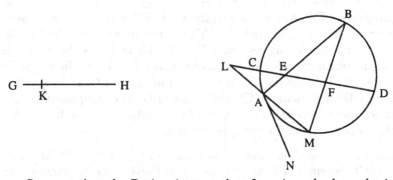

Suppose that the Porism is true; therefore since, by hypothesis, the rectangle CF, ED is to the rectangle CE, FD as the straight line HG is to the straight line GK, *dividendo*, the rectangle EF, CD will be to the rectangle CE, FD as the straight line HK is to the straight line KG (n1). Let DC be to CL as HK is to KG, and so the rectangle EF, CD will be to the rectangle EF, LC; therefore the rectangle EF, CD is to the rectangle EF, LC as the rectangle EF, CD is to the rectangle CE, FD, and consequently the rectangle CE, FD is equal to the rectangle EF, LC. Let the rectangle CE, EF be added on both sides, therefore the whole rectangle CE, ED, that is the rectangle AE, EB, is equal to the whole rectangle LE, EF; thus the points A, L, B, F lie on a circle, and consequently, AL having been joined, the angle ALE is equal to the angle ABF. But since the ratio DC to CL is given, the straight line CL will be given, for DC is given, and the point L will be given; and the point A is given, so that the straight line AL is given in position, and the angle ALE will be given, as also the angle ABF which is equal to it. Therefore since the point A is given, the straight line BF will pass through a given point on the circumference [*Dat.* 93]; let this point be M. Therefore the point M is given, which had to be shown.

It will be composed as follows.

Let it be arranged that DC is to CL as HK is to KG, let AL be joined, and let AN be drawn tangent to the circle at the point A; let the angle NAM be made equal to the angle ALC, and let AM meet the circumference again in M; then M will be the point which has to be found; that is to say, if any straight line AEB is drawn from A, meeting the circumference again in B and meeting the (n2) diameter CD in E, and if BM is joined which meets CD in F, then the rectangle CF, ED will be to the rectangle CE, FD as the straight line GH is to the straight line GK.

For since the circle touches AN and cuts AM, the angle ABM will be equal to the angle NAM [III 32], that is the angle ALC; therefore the points A, L, B, F lie on a circle. Therefore the rectangle LE, EF is equal to the rectangle (AE, EB, that is to the rectangle) CE, ED; let the rectangle FE, ED be added to both sides, therefore the whole rectangle EF, LD is equal to the whole rectangle CF, ED; and if the rectangle CE, EF is removed from both of the same rectangles LE, EF and CE, ED, the remaining rectangle LC, EF will be equal to the remaining rectangle CE, FD. Therefore as the rectangle CF, ED is to the rectangle CE, FD so (the rectangle EF, LD is to the rectangle LC, EF, and so the straight line DL is to the straight line LC, and so) GH is to GK, because DC was made to CL as HK is to KG. The composition could also be carried out by reversing the steps of the analysis, but we have given this slightly shorter version for the sake of variety.

And it is clear from this that Fermat did not set forth with sufficient generality the Theorem which he calls his third Porism; for he requires the straight line AM to be parallel to the diameter CD; but AM, as also CD, can have any desired position. Moreover, it had to be proposed as follows in the form of a Datum: Suppose that a straight line AM which is given in position is inscribed in a circle, and that straight lines AB, MB are inflected to the circumference in any manner, meeting a straight line CD which is given in position and is inscribed in the circle in the points E, F; then the ratio of the rectangle CF, ED to the rectangle CE, FD will be given. The demonstration of this is clear from the preceding composition: namely if the straight line AL is drawn to CD making angle ALC equal to the given angle ABM.

Proposition 82

"Suppose that the straight line CE is cut harmonically at the points H, M, that is to say, suppose that CE is to EM as CH is to HM, and from any point D let DE, DM, DH, DC be drawn, which are commonly called harmonicals; then any straight line KAG which is parallel to one of them CD will be bisected by the others, that is to say, KA will be equal to AG. And conversely, if a line KAG is drawn which is bisected by three of the harmonicals DH, DM, DE, then KG will be parallel to the fourth CD."

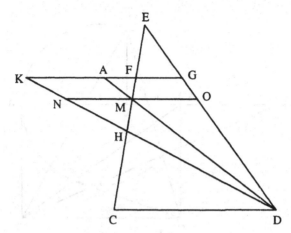

First, let KG be parallel to CD; through the point M let NMO be drawn parallel to the straight line CD; and since CH is to HM as CE is to EM, on account of the parallels CD will be to MO as CD is to NM. Therefore NM is equal to MO, consequently KA is also equal to AG.

Secondly, let KG be bisected at A; then KG will be parallel to CD. For let the straight line NMO be drawn through M parallel to the straight line CD; and therefore NM is equal to MO by the first part and KA is equal to AG; therefore, by the Lemma in note (b) to Proposition 23, KG is parallel to NO, that is to the straight line CD.

,243)

Proposition 83

(This is Fermat's fourth Proposition.)

"Let there be a circle whose diameter AB is given in position and magnitude, and on AB let two points be given equidistant from the centre; then two other points will be given with the property that, if two straight lines are inflected from them in any manner to the circumference, these will cut off two segments of the diameter adjacent to the given points, the sum of whose squares will be given."

Case 1. Here the two given points are the extremities of the diameter.

Suppose that the Porism is true, and let the points C, D be those which have to be found. Therefore if CE, DE are inflected in any manner to the circumference at E, meeting the diameter AB in the points F, G, then the sum of the squares of AG, FB is equal, by the hypothesis of the Porism, to a given area (n1). Now CA, DA have been inflected to an end of the diameter and DA cuts off no segment while CA cuts off the diameter; consequently the square of the diameter is that given area to which the sum of the squares of the segments cut off by CE, DE, or any other inflected straight lines, must be equal. Let one of the inflected straight lines CE meet the circumference again in H, let DH be joined and, having been produced, let it meet AB in K.

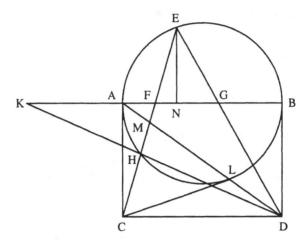

Therefore since CH, DH have been inflected from the points C, D to the circumference, meeting the diameter AB in F, K, the sum of the squares of the segments KA, FB will be equal, by hypothesis, to the given area, namely the square of AB, to which the sum of the squares of AG, FB is also equal. Therefore the square of AK is equal to the square of AG, and so the straight line AK is equal to the straight line AG. It is therefore obvious that, if from one of the points C, D, say from C, a straight line CHE is drawn in any manner to the circumference, and the joins DH, DE meet AB in K, G, then KA, AG will be equal. Whence it is clear that the straight line CA is tangent to the circle, for if it were to meet it again, the straight line drawn from the point D to this intersection would take off a certain segment of the diameter adjacent to the point A, whereas the straight line DA takes off no segment, which is impossible. Let DA meet the circumference again in L; it will be shown in the same way that the join CL is tangent to the circle. And it is shown likewise that the join DB is tangent to the circle. Therefore since CA, CL are tangent to the circle, and CHE has been drawn to the circle, meeting AL in M, then EC will be to CH as EM is to MH [Prop. 73]; therefore the four straight lines DE, DM, DH, DC are what are called harmonicals (n2). And since the straight line KAG drawn between three of them is bisected in A, the fourth DC will be parallel to KG [Prop. 82], therefore ABDC is a parallelogram. Let the perpendicular EN be drawn to AB; and since NF is to FA as EN is to AC, that is as EN is to BD, that is as NG is to GB, then NF will be to FA as NG is to GB, and *componendo*, NB is to BG as NA is to AF; therefore the rectangles AN, NB and AF, GB are similar. Consequently [VI 22], as the square of AN is to the square of AF so the rectangle AN, NB is to the rectangle AF, GB and so twice AN, NB is to twice AF, GB. But twice the rectangle AF, GB is equal to the square of FG [Converse of Prop. 44], because the squares of AG, FB are together equal to the square of AB; and, on account of the circle, the rectangle AN, NB is equal to the square of EN. Therefore twice the square of EN is to the square of FG as the square of AN is to the square of AF; and twice the square of AC is to twice the square of EN as the square of AF is to the square of FN; therefore *ex aequali in proportione perturbata*, twice the square of AC is to the square of FG as the square of AN is to the square of NF. But the square of CD, or AB, is to the square of FG as the square of CE is to the square of EF, that is as the square of AN is to the square of NF; therefore twice the square of AC is to the square of FG as the square of AB is to the square of FG. Therefore the square of AB is equal to twice the square of AC, that is to say AC is equal to the side of the square inscribed in the circle. Therefore AC is given in magnitude, but it is also given in position, so that the point C is given, and consequently the point D is given. This is what had to be shown.

It will be composed as follows.

Let the diameter AB of a circle be given in position and magnitude, and let AC, BD be drawn at right angles to it, both of which are equal to the side of the square inscribed in the circle; then C, D will be the points which have to be found, that is to say, if straight lines CE, CD are inflected from them in any manner to the circumference described on the diameter AB and they meet the diameter in the points F, G, then the squares of the segments AG, FB together will be equal to the square of the diameter AB.

For let the perpendicular EN be drawn to the diameter, and since the square of AB is equal to twice the square of AC, the square of AB will be to the square of FG as twice the square of AC is to the square of FG. But the square of AB, or of the join CD, is to the square of FG as the square of CE is to the square of EF, that is as the square of AN is to the square of NF. Therefore the square of AN is to the square of NF as twice the square of AC is to the square of FG; but twice the square of EN is to twice the square of AC as the square of NF is to the square of FA; therefore *ex aequali in proportione perturbata*, twice the square of EN is to the square of FG as the square of AN is to the square of AF. Now, as shown in the analysis, the square of AN is to the square of AF as twice the rectangle AN, NB, that is twice the square of EN, is to twice the rectangle AF, GB. Therefore twice the square of EN is to the square of FG as twice the square of EN is to twice the rectangle AF, GB. Therefore the square of FG is equal to twice the rectangle AF, GB, and consequently the squares of AG, FB are together equal to the square of AB [Prop. 44]. Q.E.D.

This demonstration which is contained in the Composition is given in a different way by the most distinguished Euler in the new *Acta Petropolitana* for the year 1750, p. 49 of volume I. However he does not investigate the Construction at all but assumes it as Fermat had given it. And indeed I first demonstrated Fermat's Proposition in this Case in the year 1715, before I had learned anything about the nature of the Porisms (n3).

Case 2. Here the two given points are inside or outside the circle.

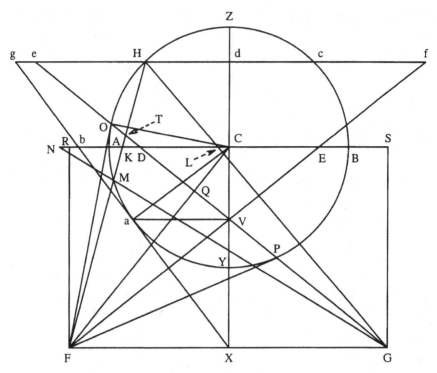

Let there be a circle whose diameter AB is given in position and magnitude, and let its centre be C; and let there be given on the diameter either inside or outside the circle two points D, E which are equidistant from the centre. Two other points will be given with the property that if two straight lines are inflected from them to the circumference in any manner, these will cut off two segments of the diameter adjacent to the given points D, E, the sum of whose squares will be given.

Suppose that the Porism is true; and let F, G be the two points which have to be found. Therefore if two straight lines FH, GH are inflected from them to the circumference in any manner, meeting the diameter AB in K, L, by hypothesis, the sum of the squares of the segments cut off KE, LD will be given. Let FH meet the circumference again in M, and let GM be joined, meeting the diameter in N; therefore again, since FM, GM have been inflected from the points F, G to the circumference, cutting off the segments KE, ND adjacent to the points E, D, the sum of the squares of KE, ND will be given, and consequently it will be equal to the sum of the squares of KE, LD. Therefore the square of ND will be equal to the square of DL, and the straight line ND will be equal to the straight line DL. It follows therefore that the points F, G have the following property: if from one of them F any straight line is drawn which meets the circumference in H, M, the straight lines GH, GM drawn from the other will cut off from the diameter equal segments LD, DN adjacent to the given point D. Let GD be joined and let it

meet the circumference in O, P, and let FO, FP be joined; therefore since FO
has been drawn to the circumference at O, and the straight line GO which
passes through the point D takes off no segment adjacent to the same point,
the straight line FO will meet the circumference nowhere other than in O,
that is it will be tangent to the circle; it will be shown similarly that the
straight line FP is tangent to the circle. Let FC be joined and let it meet
the straight line OP in Q, and since FO, FP are tangent to the circle, FQ
will be perpendicular to OP. Let the perpendicular FR be drawn to AB, and
since angles FRD, FQD are right angles, the points F, R, D, Q will lie on
a circle, therefore the rectangle RC, CD is equal to (the rectangle FC, CQ,
that is to) the square of the join CO, or of CA (n4). Therefore CD, CA,
CR are proportional, and CD, CA are given, so that CR is also given; and
the point C is given, consequently the point R will also be given, and the
straight line RF will be given in position. Similarly if FE is joined, and the
perpendicular GS is drawn to AB, the straight lines CE, CB, CS will be
shown to be proportional. Therefore the straight lines CR, CS are equal to
each other, and consequently the point S is given.

Let the straight line FH meet OP in T, and since FO, FP are tangent
to the circle, FH will be to FM as HT is to TM [Prop. 73]; therefore, FG
having been joined, the straight lines GF, GM, GT, GH will be harmonicals,
and the straight line NDL is bisected by three of them GN, GD, GL in D,
consequently the same straight line NL is parallel to the fourth FG [Prop.
82]; therefore FRSG is a parallelogram.

Let FE, GD meet each other in V, and let CV be joined, meeting FG in
X. Therefore since DC, CE are equal, and DE, FG are parallel, FX, XG will
be equal to each other; therefore FX is equal to RC, since they are halves
of the equal quantities FG, RS, and they are parallel, consequently CX is
parallel to RF, and hence the angle FXC is a right angle. Let CV meet the
circumference in Y, Z; and since the points F, Q, V, X lie on a circle, for the
angles FQV, FXV are right angles, the rectangle XC, CV will be equal to the
rectangle FC, CQ, that is the square of CO, or CY; therefore CV, CY, CX
are proportional. And since the angles FCX, VGX are equal, the right-angled
triangles FXC, VXG will be equiangular to each other; therefore VX is to
XG as FX is to XC, consequently the rectangle FX, XG, that is the square of
FX, is equal to the rectangle CX, XV; and the square of CY, or CA, is equal
to the rectangle XC, CV, therefore the squares of the given straight lines FX,
CA, or RC, CA, are equal to the rectangles CX, XV, XC, CV, whose sum is
the square of CX, which consequently is given. Therefore the straight line
CX is given, so that the point X is given as also are the points F, G. Through
the point V let the straight line aV be drawn parallel to AB, and let it meet
the circumference in a; let Ca be joined as also Xa which will be tangent to
the circle at a, because CV, CY, CX are proportional; therefore the angle
CaX is a right angle, consequently the square of aX is equal to the rectangle
CX, XV, that is to the square of FX. Therefore the straight line aX is equal

to the straight line FX. Let Xa meet the straight line AB in b, and since, on account of the parallels, FX is to CE as (XV is to VC, that is as) Xa is to ab, and FX is equal to Xa, then CE will be equal to the straight line ab. Through H let Hd be drawn parallel to FG, and let it meet the circumference again in c, and the straight lines XZ, GD, FE, Xa in the points d, e, f, g. And since FX is to df as (XV is to Vd, that is as) Xa is to ag, and FX, Xa have been shown to be equal, df will be equal to ag. Moreover since ag is tangent to the circle, the square of ag, that is df, will be equal to the rectangle Hg, gc. Let the square of Hd be added on both sides, and the squares of Hd, df together will be equal to the square of gd. Now ef is bisected in d, because FG is bisected in X, consequently the squares of eH, Hf are together twice the squares of Hd, df together, that is (twice) the square of gd. Now, on account of the parallels, eH is to DL as (HG is to GL, that is as HF is to FK, that is as) Hf is to KE; and their squares are proportional; therefore as the sum of the squares of eH, Hf is to the sum of the squares of DL, KE so (the square of eH is to the square of DL, and so the square of HG is to the square of GL and so the square of dX is to the square of CX, and so) the square of gd is to the square of bC. But the sum of the squares of eH, Hf has been shown to be twice the square of gd, therefore the sum of the squares of DL, KE is twice the square of bC, that is twice the sum of the squares of Ca, ab, or CA, CE, or CA, CD. Moreover the straight lines CA, CD are given, therefore the sum of the squares of DL, KE is given. It has therefore been shown that the two points F, G with the following property are given: straight lines FH, GH inflected from them in any manner to the circumference cut off segments KE, LD adjacent to the given points D, E, which make the sum of their squares given. These are what had to be found.

It will be composed as follows.

Let the diameter AB of a circle be given in position and magnitude, and on it let two points D, E be given equidistant from the centre C either inside or outside the circle. Let a third proportional CR be found for the straight lines CD, CA, and let CS be equal to CR and be located on the diameter on the other side of the centre. Let the straight line CZ be drawn to the circumference at right angles to RC, and let RZ be joined; let RF be drawn parallel to ZC and equal to RZ (n5), and let the parallelogram FRSG be completed; the points F, G will be those which had to be found, that is to say, if straight lines FH, GH are inflected from them to the circumference in any manner, meeting the diameter AB in K, L, the squares of the segments KE, LD adjacent to the points E, D will together be equal to a given area, namely twice the sum of the squares of CA, CD, or the sum of the squares of AD, DB. And this is the same construction that Fermat gives on p. 118 of his works.

For let ZC meet the straight line FG in X and the circumference again in Y, and let Xa be drawn tangent to the circle at a, and let it meet the diameter AB in b; then through H let a straight line be drawn parallel to

AB, meeting the circumference again in c, and the straight lines XZ, Xa in d, g. Further let the join GD meet XC in V, and since XV is to VC as (XG is to DC, that is as) FX is to CE, the points F, V, E will be in a straight line; let it be drawn, and let it meet the straight line Hc in f; let GD meet the same Hc in e. Finally let aC, aV be joined.

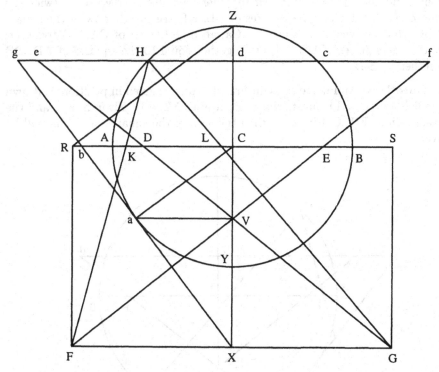

Therefore since Xa is tangent to the circle, the angle XaC will be a right angle; consequently the squares of Ca, aX together will be equal to (the square of CX, or RF, that is to the square of RZ, that is to) the squares of ZC, CR together, of which the squares of Ca, CZ are equal; therefore the remaining square of aX is equal to the remaining square of CR; therefore the straight line aX is equal to the straight line CR, that is to the straight line FX. And since CR is to CA as CA is to CD, that is Xa is to aC as aC is to CE, and Xa is to aC as aC is to ab, then ab will be equal to CE. And since XV is to VC as (FX is to CE, that is as) Xa is to ab, then aV will be parallel to Cb. But FX is to df as (XV is to Vd, that is as) Xa is to ag, and FX, Xa have been shown to be equal, consequently df, ag are equal. Therefore the square of df is equal to the square of ag, that is to the rectangle Hg, gc, because ag is tangent to the circle; let the square of Hd be added to both sides, and the squares of Hd, df together will be equal to the square of gd. But the squares of eH, Hf together are twice the squares of Hd, df together, because ef is bisected in d, for FG is bisected in X, therefore the same squares are twice the square of

gd (n6). Now, because of the parallels, the square of eH is to the square of
DL as the square of Hf is to the square of KE; therefore as the squares of eH,
Hf together are to the squares of DL, KE together so (the square of eH is to
the square of DL, and so the square of eG is to the square of GD, and so the
square of dX is to the square of XC, and so) the square of gd is to the square
of bC; and the squares of eH, Hf together have been shown to be twice the
square of gd, therefore the squares of DL, KE are together twice the square
of bC, that is twice the squares of Ca, ab together, or of CA, CD together;
that is the squares of DL, KE are together equal to the squares of AD, DB
together. Q.E.D.

Corollary. And if the straight line kl, which is given in position, is drawn
parallel to AB, and it meets the straight lines XZ, GD, FE in h, k, l, and the
straight lines FH, GH in m, n, then the sum of the squares of kn, ml will be
given.

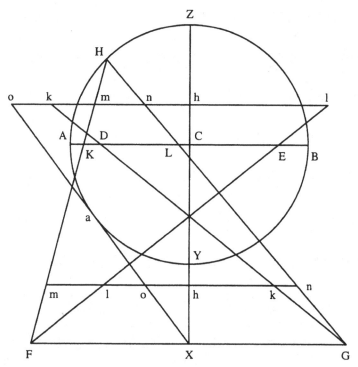

For let Xa meet kl in o; and since Xa, kl are given in position, the point
o will be given, and h is given, therefore the straight line oh is given in
magnitude. But, on account of the parallels, the square of kn is to the square
of DL as the square of ml is to the square of KE, therefore the sum of the
squares of kn, ml is to the sum of the squares of DL, KE as (the square of
kn is to the square of DL, that is as the square of nG is to the square of GL,
that is as the square of hX is to the square of CX, that is as) the square of

ho is to the square of CB. And the sum of the squares of DL, KE is equal to twice the square of BC, therefore the sum of the squares of kn, ml is equal to twice the square of oh. But oh is given, consequently the sum of the squares of kn, ml is also given (n7).

Proposition 84

245)

(This expresses the preceding proposition more generally.)

"Suppose that the circle AZB whose centre is C and the straight line kl to which the straight line Ch from the centre is perpendicular have been given in position, and that the two points k, l on kl are given and are equidistant from the point h. Then two other points will be given with the property that, if two straight lines are inflected from them in any manner to the circumference, these will cut off from the straight line kl segments adjacent to the points k, l the sum of whose squares will be given."

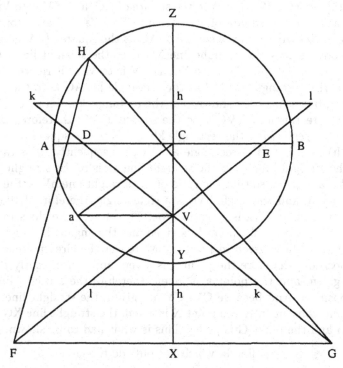

Let the diameter AB parallel to kl be drawn; therefore if two points F, G are found and if on the diameter AB two points D, E equidistant from the centre are shown to be given such that straight lines inflected in any manner from the points F, G to the circumference will cut off from the diameter AB segments adjacent to the points D, E the sum of whose squares will be given and such that the join GD passes through the given point k and the join FE passes through the given point l, it is clear from the Corollary of the preceding Proposition that the same inflected lines will cut off from the straight line kl two segments adjacent to the points k, l the sum of whose squares will be given.

Let Ch meet the circumference in Y, Z; it is therefore assured by the previous Proposition (n1) that a straight line kD may be drawn from the given point k meeting the straight lines AB, YZ in D, V, such that if CX, the third proportional to CV, CY, is taken and FX is drawn through X parallel to AB and meeting kD in G, then GX, AC, CD are also proportional.

Let this be done; therefore, on account of the parallels, hV is to VC as kh is to DC, so that the rectangle kh, XG is to the rectangle DC, XG, that is, by hypothesis, to the square of AC, as hV is to VC, that is as the rectangle hV, VC is to the square of VC. Let Va be drawn to the circumference parallel to AB; and since the angle CVa is a right angle and XC, CY, or Ca, and CV are proportional, the square of Va will be equal to the rectangle XV, VC (n2). And since XV is to VC as (XG is to DC, that is as) the square of XG is to the rectangle XG, DC, that is to the square of AC, and XV is to VC as the square of aV is to the square of VC, because XV, Va, VC are proportional, the square of XG will be to the square of AC as the square of aV is to that of VC, and consequently the straight line XG is to the straight line AC as aV is to VC, and, *invertendo*, AC is to XG as CV is to Va. Therefore the square of CV is to the rectangle aV, VC as the rectangle kh, AC is to the rectangle kh, XG (n3); but it has been shown that the rectangle kh, XG is to the square of AC as the rectangle hV, VC is to the square of VC; therefore, *ex aequali in proportione perturbata*, the rectangle kh, AC is to the square of AC as the rectangle hV, VC is to the rectangle aV, VC; consequently the straight line hV is to the straight line Va as the straight line kh is to the straight line AC. And kh, AC are given, so that the ratio of the straight line hV to the straight line Va is given; and the angle hVa is a right angle, therefore if the join ha is drawn, the triangle hVa is given in type. Therefore, since the straight line hV is given in position, the point h is given and the angle Vha is given, then the straight line ha is given in position. Moreover the circumference is also given in position; therefore the point a is given, and consequently the point V is also given, and the point k is given; therefore the straight line kV is given in position. And because CV, CY are given, the straight line CX will be given, and consequently the point X is given, the straight line XG is given in position and the point G is given. This is what had to be shown.

See the line lk which lies outside the circle.

However, since it is required for the composition that a straight line hp equal to hk be located on hC and pq be drawn parallel and equal to AC and that the join hq meet the circumference in a certain point a, for thus, if the straight line aV is drawn parallel to the straight line qp to meet hC, then hV will have the same ratio to Va as hp has to pq, that is as kh has to CA, it is obvious that the problem cannot be solved if hq does not meet the circumference.

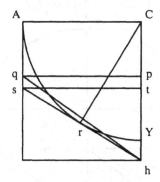

Now in the case where the point h is inside the circle, hq will always meet the circumference; but if h is outside the circle, let a straight line hr be drawn from it tangent to the circle in r; therefore every straight line which falls on the circle side of the tangent hr will meet the circle, while one falling on the other side will not meet the circle. In order that it fall on the circle side it is necessary that the straight line kh be greater than the tangent hr; this is shown as follows. Let As be drawn parallel to hC, and let the tangent hr meet it in s, and let the straight line st be drawn parallel to the straight line AC, or qp, to meet hC, and let Cr be joined. Therefore since in triangles sth, Crh the angles sth, Crh are right angles, the angle ths is common and the equal sides st, Cr are opposite it, then th will be equal to hr. But if any straight line is drawn from the point h to meet the circle, it will necessarily meet the straight line sA above the point s; moreover the straight line hq, which, by hypothesis, meets the circumference, cuts the straight line sA in q, and consequently the straight line qp, which is parallel to the straight line st, will fall above it; therefore hp is greater than ht, that is kh is greater than the tangent hr. These things having been set forth,

The Problem will be composed as follows.

See the straight line kl which is outside the circle.

Let the straight line kl be given in position, and let the points k, l be given; from the centre let the straight line Ch be drawn perpendicular to kl, so that, by hypothesis, lh, hk are equal and, in the case where kl does not meet the circle, let kh be greater than the straight line hr which is tangent to the circle in r. Moreover, let the straight line hp equal to the straight line kh be located on hC, and let pq be drawn parallel and equal to AC, so that, as has been shown, the join hq will meet the circumference. Let a be the second point of intersection, and let aV be drawn parallel to AC to meet hC; let CX, the third proportional to CV, CY, be found, and through X let the straight line FX be drawn parallel to the straight line AB; let the join Vk meet FX in G and let it meet AB in D; then the straight lines GX, AC, CD will be proportional.

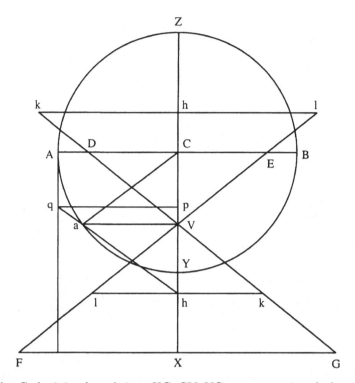

For let Ca be joined, and since XC, CY, VC are proportional, the rectangle XC, CV will be equal to the square of CY, or Ca; and, when the square of CV has been taken away from both sides, the rectangle XV, VC will be equal to the square of Va. And since hp is to pq, that is hk is to AC, as hV is to Va, *permutando*, kh will be to hV as AC is to aV; but DC is to CV as kh is to hV; therefore DC is to CV as AC is to aV, and since the rectangles XG, DC and XV, VC upon DC, CV are similar (n4), the square of AC will be to the square of aV as the rectangle XG, DC is to the rectangle XV, VC [VI 22]. But the square of aV is equal to the rectangle XV, VC, therefore the square of AC is equal to the rectangle XG, DC, and consequently the straight lines XG, AC, CD are proportional. Q.E.D.

Besides since it was shown in the analysis that the square of GX is to the square of AC as the square of aV is to that of VC, *componendo*, the square of GX along with that of AC will be to the square of AC as the squares of aV, VC together, that is the square of AC, is to the square of VC. And the square of CX is to the square of CY, or AC, as the square of AC is to that of VC; therefore the squares of XG, AC together are equal to the square of CX. Therefore if CR is made equal to XG, and the straight line RF is drawn parallel to the straight line CX to meet XG, the square of RF, that is of CX, will be equal to the squares of RC, CA, and consequently, CE having been made equal to DC, by Case 2 of Proposition 83, F, G are the points with the property that if straight lines FH, GH are inflected from them to the circumference in any manner cutting off from the diameter segments KE, DL adjacent to the points D, E, then the squares of KE, DL will together be equal to a given area, namely the squares of AD, DB; and by the Corollary of Proposition 83 the same straight lines FH, GH will cut off from the straight line kl segments ml, kn adjacent to the given points k, l, whose sum of squares will be equal to a given area, namely twice the square of oh. This is what had to be shown.

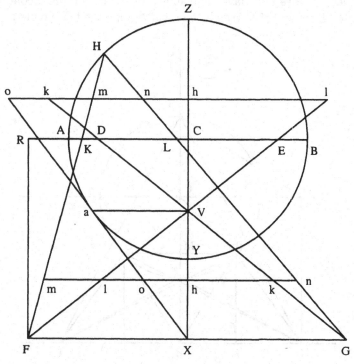

Mr Matthew Stewart has expressed the same Proposition of Fermat which is contained in Case 2 of Proposition 83 still more generally in the following Proposition, which he communicated to me many years ago, viz. (see the figure for Case 2 of Proposition 83)

Suppose that the diameter AB of a circle, whose centre is C, is given in position and magnitude; and let two points D, I be given on AB which are located on opposite sides of the centre and at unequal distances from it; two other points will be given such that, if two straight lines are inflected from them to the circumference in any manner, one of them meeting the diameter in K, the other meeting it in L, then the square of KI along with the area to which the square of DL has the same ratio as the straight line DC has to the straight line CI will be given. Moreover this Proposition can be investigated and proved by a method not much different from that which was used in Case 2 of Proposition 83.

(p.245)

Proposition 85

(Fermat's fifth Proposition follows immediately from this)

"Suppose that a circle has been given in position and magnitude and that a straight line has been given in position; two points will be given on a straight line which is parallel to this one such that, if two straight lines are inflected from them to the circumference in any manner, then the sum of the squares of the inflected lines will have the same ratio to the triangle contained by the inflected lines and the straight line which joins the two points as a given ratio."

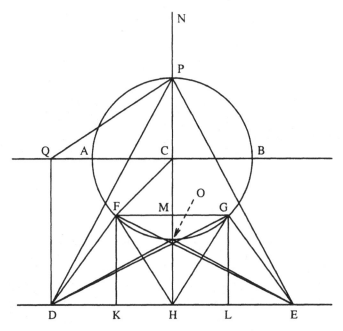

Let the circle be given in position whose diameter AB is given in position and whose centre is C. And suppose that the Porism is true, so that two points D, E are given in a straight line which is parallel to AB such that, if DF, EF are inflected from them to the circumference in any manner, then the sum of the squares of DF, EF will have the same ratio to the triangle contained by DF, FE and the join DE as the given ratio which a has to b. Let FG be drawn through the point F parallel to DE, and let it meet the circumference again in G; further let DE be bisected in H, and let perpendiculars FK, GL be drawn to DE.

Therefore since, by hypothesis, the sum of the squares of DF, FE has to triangle DFE the ratio which a has to b, and since likewise the sum of the squares of the joins DG, GE has the same ratio to the triangle DGE, and moreover triangle DFE is equal to triangle DGE [I 37], the sum of the squares of DF, FE will therefore be equal to the sum of the squares of DG,

GE. Now the sum of the squares of DF, FE is equal to the squares of DK, KE along with twice the square of FK [I 47], that is to twice the squares of DH, HK, KF [II 9], that is to twice those of DH, HF (n1). And likewise the sum of the squares of DG, GE is equal to twice the squares of DH, HG. Therefore the square of HF is equal to the square of HG, and the straight line HF is equal to the straight line HG. Let FG be bisected in M, and let HM be joined; the angle FMH is therefore a right angle [I 8], and since HM bisects the straight line FG in the circle and cuts it at right angles, the straight line HM will pass through the centre C [Cor. to III 1]. Let HC be produced to N such that two times HN is to HD as (a is to b, that is as the sum of the squares of DF, FE is to triangle DFE, that is as) twice the squares of DH, HF is to triangle DFE. Therefore the sum of the squares of DH, HF will be to (twice the triangle DFE, that is to) the rectangle DE, FK as (the straight line NH is to the straight line DE, that is as) the rectangle NH, FK is to the rectangle DE, FK. Consequently the squares of DH, HF together are equal to the rectangle NH, FK, that is to the rectangle NH, HM (n2). Let the square of HM be taken away from both sides, and the squares of DH, FM together will be equal to the rectangle HM, MN. Let HN meet the circumference in the points O, P, and, if DO, EO are inflected, it will be shown, as in what has gone before, that the square of DH is equal to the rectangle HO, ON. Likewise, if DP, EP are inflected, the square of DH will be equal to the rectangle HP, PN. Therefore the rectangle HO, ON is equal to the rectangle HP, PN, so that the straight line HO is also equal to the straight line PN (n3); and OC is equal to the straight line CP, and consequently HC is equal to CN.

Let DQ be drawn parallel to HC to meet AB, and let FC be joined. And since it has been shown that the squares of DH, FM together are equal to the rectangle HM, MN, let the square of MC be added to both sides, and the squares of DH, FC together will be equal to the square of HC (n4). Now the squares of DH, FC are equal to the squares of QC, CP, that is the square of the join PQ, which consequently will be equal to the square of HC, and the straight line PQ will be equal to the straight line HC. But since the ratio a to b is given, that is the ratio which twice HN has to HD, the ratio which HC, a fourth part of twice HN, has to HD, that is the ratio which PQ has to QC, will be given. Therefore the right-angled triangle CPQ is given in type [Dat. 46], and CP is given in magnitude, therefore PQ, QC, that is CH, HD will be given in magnitude. But CH is also given in position, having been drawn from the given point C at right angles to AB which is given in position; therefore the point H is given, and the straight line HD is given in position; and HD, DE are given in magnitude, so that the points D, E are given. These things are what were required to be found.

It will be composed as follows.

Let the diameter AB of a circle be given in position and magnitude, and let the centre be C. Moreover let the ratio which the sum of the squares must have to the triangle be that which four times the given straight line RS has

to the given straight line RC; and let the points which require to be found
be on a straight line which is parallel to the straight line AB. And since PQ
must be to QC as RS is to RC, and PQ is greater than QC, then RS must be
greater than RC. Let this be the case, and let CP be drawn at right angles to
AB, and from the point R let RT equal to RS be drawn to it; now from the
point P let PQ be drawn parallel to RT to meet AB, and from the point Q
let QD equal to the straight line QP be drawn at right angles to QB; finally
let DH be drawn parallel to QC to meet PC, and let it be produced to E so
that HE is equal to HD; then D, E will be the points which had to be found.

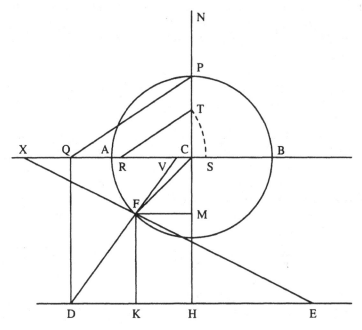

For from the points D, E let straight lines DE, EF be inflected to the
circumference in any manner, and let FM be drawn parallel to DH to meet
CH, and let HC be produced to N so that HC, CN are equal, and let CF be
joined. And since the square of CH, or PQ, is equal to (the squares of QC, CP,
that is to the squares of DH, FC, that is to) the squares of DH, FM, MC, let
the square of MC be taken away from both sides, and the rectangle HM, MN
will be equal to the squares of DH, FM, or of DH, HK; let the square of HM,
or KF, be added to both sides, and the rectangle NH, HM will be equal to
the three squares of DH, HK, KF. Therefore these three squares are to the
rectangle DE, KF as (the rectangle NH, HM is to the same rectangle DE, KF,
that is as) the straight line NH is to the straight line DE. Therefore two times
the squares of DH, HK, KF is to the rectangle DE, KF as twice the straight
line NH is to DE; and two times the same squares, that is the squares of DF,
FE together, are to the triangle DFE, namely half the rectangle DE, KF, as
twice the straight line NH is to DH, which is half the straight line DE, that

is as four times CH, or four times PQ, is to QC, that is as four times RT, or RS, is to RC, that is in the given ratio. Q.E.D.

Corollary. And if DF, EF are produced to meet AB (or any line parallel to it) in V, X, it is clear that the sum of the squares of VF, FX will be to the triangle VFX as the sum of the squares of DF, FE is to the triangle DFE, that is in the given ratio. And this is Fermat's fifth Proposition.

Proposition 86

"Let the straight line AB be cut harmonically in C, D, that is to say BA is to AC as BD is to DC; and from any point E let EA, EC, ED, EB be drawn; a straight line GF meeting them in any manner will be cut harmonically in the points of intersection."

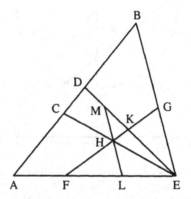

Let GF meet the straight lines EA, EC, ED, EB in the points F, H, K, G, and through H let LHM be drawn parallel to EB, and let it meet the straight lines EA, ED in L, M. Therefore, on account of the parallel lines, GF is to FH as EG is to LH, that is as EG is to HM, because LM is bisected in H [Prop. 82]; moreover, EG is to HM as GK is to KH; therefore GK is to KH as GF is to FH.

Proposition 87

"Let there be a circle whose diameter is OCP, and on this diameter let two points K, N be taken on the same side of the centre C, which make CK, CO, CN proportional; if through one or other of the points K, N, say N, a straight line ND is drawn at right angles to the straight line CN, and through the other point K a straight line KE is drawn in any manner to meet ND in E, and the circumference in the points Q, R, then QE will be to ER as QK is to KR."

Case 1. Here the point E is outside the circle. Let the straight line EC be drawn from the point E to the centre C, and let the perpendicular KI be drawn from the point K to EC; therefore since the points E, N, K, I are in a circle, because the angles at N, I are right angles, the rectangle EC, CI will be equal to (the rectangle NC, CK, that is to) the square of the semidiameter CO; consequently the straight lines drawn from the point E to the points in which KI meets the circumference will be tangent to the circle. Therefore the straight line EQ which meets the circumference in Q, R and the chord of contact in K, will be cut in the same points in such a way that QE is to ER as QK is to KR [Prop. 73].

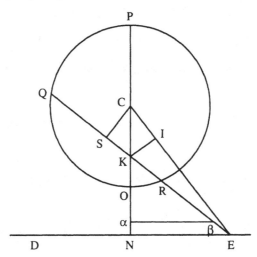

Case 2. Now the point E is inside the circle. Since CK, CO, CN are proportional, and EN is perpendicular to CK, the straight lines which are drawn from the point K to the points in which EN meets the circumference, will be tangent to the circle, and the straight line KQ meets the straight line which joins the points of contact in E, therefore QK is to KR as QE is to ER [Prop. 73].

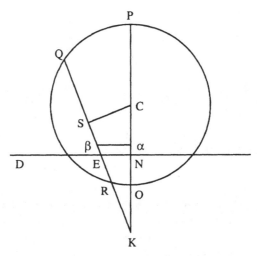

Conversely, if on the diameter OCP of the circle two points K, N are taken on the same side of the centre, and if NE is drawn through one of them at right angles to the diameter, and through the other point K a straight line KE is drawn in any manner to meet the straight line NE in E, and the circumference in the points Q, R, and if QE is to ER as QK is to KR, then CK, CO, CN will be proportional.

For if this is not the case, let CK, CO, Cα be proportional and through α let the straight line αβ be drawn parallel to NE to meet EK; therefore, by this Proposition, as Qβ is to βR so (QK will be to KR, that is, by hypothesis, as) QE will be to ER, which is impossible.

This may be shown directly as follows. From the centre C let the perpendicular CS be drawn to QR; this will bisect QR. Therefore since QK is to KR as QE is to ER, and QR is bisected in S, by Case 1 of Proposition 64 the rectangle ES, SK is equal to the square of SR. But the points C, S, N, E are on a circle because the angles at S, N are right angles, and so the rectangle SK, KE is equal to the rectangle CK, KN; and the squares of CS, SK together are equal to the square of CK. Therefore (n) the sum, if the point E is outside the circle, or the difference, if it is inside, namely the square of CS along with the rectangle ES, SK, that is along with the square of SR, that is the square of the semidiameter CR, is equal to the whole or to the remaining rectangle, namely KC, CN. Therefore CK, CO, or CR, and CN are proportional.

246)

Proposition 88

"Let there be a circle whose diameter is OCP, and let two points K and N be taken on this diameter on the same side of the centre C which make CK, CO, CN proportional, and, ND having been drawn from the point N at right angles to CN, let two points D, E be taken on the straight line ND on opposite sides of N if the point N is outside the circle but on the same side if it is inside, and let the rectangle DN, NE be equal to the rectangle ON, NP; then from either of the points D, E, say E, if both points are outside the circle, but from the point E which is inside the circle if one is outside and the other inside the circle, let the straight line EK be drawn to meet the circumference in Q, R; the joins DQ, DR will be tangent to the circle."

Let DC be joined, meeting QR in S, and since CK, CO, CN are proportional, the rectangle NC, CK will be equal to the square of CO; let these equal quantities be subtracted from the square of CN, if N is outside the circle, but if it is inside, let the square of CN be subtracted from them; in both cases, the rectangle CN, NK will be equal to the rectangle (ON, NP, that is, by hypothesis, to the rectangle) DN, NE (n). Therefore DN is to NC as KN is to NE, and they are about equal angles DNC, KNE, for both are right angles; therefore the triangles DNC, KNE are equiangular [VI 6], and the angle CDN is equal to the angle NKE, or to the angle CKS; and in triangles DCN, KCS the angle DCN is common, therefore the angle KSC is equal to the right angle DNC; consequently the points D, N, K, S are on a circle [Converse of III 22 or III 21]. Therefore the rectangle DC, CS is equal to the rectangle NC, CK, that is to the square of the semidiameter CO, and consequently the straight lines DQ, DR are tangent to the circle, for the angle CSK is a right angle.

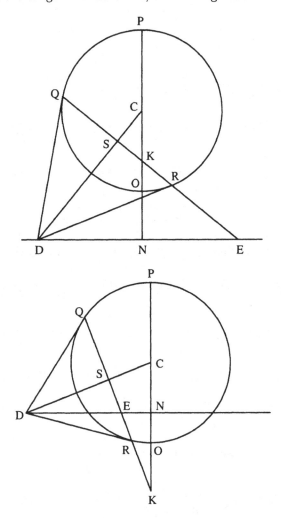

(p.246)

Proposition 89

"Let there be a straight line AB, and two points D, E not on it, and suppose that if from either of them, say D, a straight line DGF is drawn in any manner to the circle OGP whose diameter is OP, the joins EG, EF will cut off equal segments of the straight line AB adjacent to the point K, the foot of the perpendicular CK drawn from the centre C to the straight line AB, then the join DE will be parallel to the straight line AB; moreover if CK meets DE in N and the circumference in O, P, then CK, CO, CN will be proportional; and the rectangle DN, NE will be equal to the rectangle ON, NP. Suppose conversely that CK, CO, CN are proportional, and CKN is perpendicular to AB, DE, and the rectangle DN, NE is equal to the rectangle ON, NP; if from either of the points D, E, [say D], a straight line DGF is drawn in any manner to the circle, and EG, EF are joined, then these will cut off equal segments of the straight line AB adjacent to the point K."

Part 1. Let EK be joined and let it meet the circumference in Q, R, and let DQ, DR be joined; therefore since DQ has been drawn from the point D to the circle, and the join EQ takes off no segment adjacent to the point K, for it passes through the point K, the straight line DQ will be tangent to the circle;

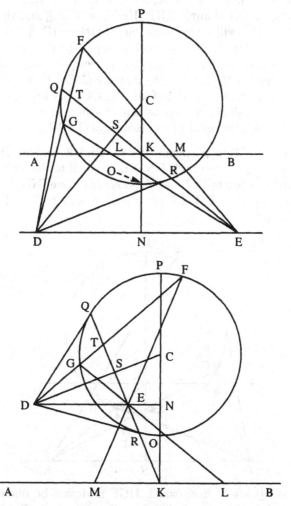

for if it were to meet it again, the straight line drawn from the point E to this intersection would take off some segment adjacent to the point K, but it takes off none, because the straight line EQ takes off none, therefore the straight line DQ is tangent to the circle (n1). Let DC be joined and let it meet QR in S, and let DF meet the same straight line in T, and let EG, EF meet AB in L, M; therefore since the straight lines DQ, DR are tangent to the circle, and the straight line DF meets the straight line QR in T, then FD will be to DG as FT is to TG [Prop. 73]; therefore ED, EG, ET, EF are harmonicals, and since the straight line LKM is bisected by three of them EG, ET, EF in K, by hypothesis, LM, or AB, will be parallel to the fourth ED [Prop. 82]. Now the straight line DC drawn to the centre from the point of intersection of the tangents DQ, DR is at right angles to the straight line QR, and the angle DNK is a right angle, for DN has been shown to be parallel to AB; therefore the points D, N, K, S are on a circle; consequently (n2) the angle CDN is equal to the angle NKE [III 22 or III 21], and the right-angled triangles DNC, KNE will be equiangular; therefore KN is to NE as DN is to NC, and the rectangle DN, NE will be equal to the rectangle CN, NK. Now since the points D, N, K, S are on a circle, the rectangle NC, CK is equal to [Corollary to III 36] (the rectangle DC, CS, that is to) the square of the semidiameter CO (n3); therefore CK, CO, CN are proportional; and consequently, as was shown in Proposition 88, the rectangle CN, NK, that is the rectangle DN, NE, is equal to the rectangle ON, NP.

Part 2. Now let CK be perpendicular to AB, DE, and let CK, CO, CN be proportional, and let the rectangle DN, NE be equal to the rectangle ON, NP; if from either of the points D, E, say D, a straight line DGF is drawn in any manner to the circle, and EG, EF are joined, meeting AB in L, M, then LK will be equal to KM.

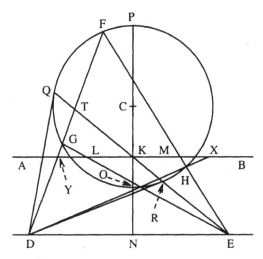

First, let the point D from which DGF is drawn be outside the circle, and let EK be joined to meet the circumference in Q, R; and since CK, CO, CN are proportional, and DN is perpendicular to CN, and the rectangle DN, NE is equal to the rectangle ON, NP, the joins DQ, DR will be tangent to the circle [Prop. 88]. Let DF meet the straight line QR in T; therefore the straight line DF is cut harmonically in the points D, G, T, F [Prop. 73], and the straight lines ED, EG, ET, EF are harmonicals. Now the straight line LM is parallel to one of them ED, and so it will be bisected by the other three which meet it in the points L, K, M, that is LK, KM are equal. And if a straight line EHF is drawn to the circle from the point E, if it is outside the circle, and DF, DH are joined to meet AB in Y, X, it will be shown likewise that YK is equal to KX.

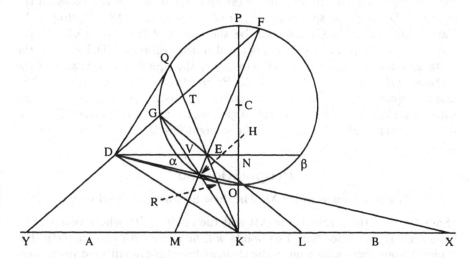

Now let the point E be inside the circle, and through it let a straight line HEF be drawn in any manner, meeting the circle in H, F, and let the joins DH, DF meet the straight line AB in X, Y; again in this case KX will be equal to KY. Let DF meet the circumference again in G, and let the join KE meet DF in T, while the join KG meets DE in V. Therefore ED, EG, ET, EF are harmonicals, as has been shown; and since GK has been drawn between them, it will be cut harmonically in the points V, G, K in which it meets three of them ED, EG, ET and in the point in which it meets the fourth EF [Prop. 86]. But the same straight line GK is cut harmonically in the points V, G, K, and the point in which it meets the circumference again [Prop. 87]; therefore this fourth point is on the straight line EF and on the circumference; therefore it will be at their intersection H; therefore GK passes through the point H, and consequently DG, DV, DH and the join DK are harmonicals. But the straight line YX is parallel to one of them DV; therefore it will be bisected by the other three in the points Y, K, X, that is YK is equal to KX.

Corollary. Hence, since it has been established, from the hypothesis of Part 1, that the rectangle DN, NE is equal to the rectangle ON, NP, the rectangle GD, DF will be to the rectangle HE, EF as the straight line DN is to the straight line NE.

For if the straight line DE is outside the circle, the rectangle GD, DF is equal [f] to the square of DN along with the rectangle ON, NP, that is by Part 1 of this Proposition to the square of DN and the rectangle DN, NE, that is to the rectangle ED, DN; it will be shown similarly that the rectangle HE, EF is equal to the rectangle DE, EN. Therefore the rectangle GD, DF is to the rectangle HE, EF as the rectangle ED, DN is to the rectangle DE, EN, that is as the straight line DN is to the straight line NE.

Now [*] (n4) if the straight line DE meets the circle in the points α, β, the rectangle GD, DF, that is the rectangle αD, Dβ, will be the excess of the square of DN and the square of αN, or of the rectangle ON, NP, that is, by Part 1, the rectangle GD, DF will be the excess of the square of DN and the rectangle DN, NE; this excess is equal to the rectangle ND, DE. Now the rectangle HE, EF, or αE, Eβ, is the excess of the square of αN, that is of the rectangle ON, NP, or, by Part 1 of this Proposition, of the rectangle DN, NE, and the square of EN; this excess is equal to the rectangle DE, EN. Therefore the rectangle GD, DF is to the rectangle HE, EF as the rectangle ND, DE is to the rectangle DE, EN, that is as the straight line DN is to the straight line NE.

(p.247)

Proposition 90

(This is a Porism which Mr Matthew Stewart proposed to me.)

"Suppose that the straight line AB and the circle OGP, whose centre is C, have been given in position; two points will be given with the property that if two straight lines which meet the straight line AB are inflected from them to the circumference in any manner, then the rectangle contained by the segments of one of them between the straight line AB and the circumference will have the same ratio to the rectangle contained by the segments of the other between the same straight line and circumference as the given ratio which a given straight line a has to a given straight line b."

Suppose that the Porism is true, and that the points D, E are those which are asserted to be given. Therefore if straight lines DF, EF have been inflected to the circumference in any manner, meeting the circumference again in G, H and the straight line AB in the points Y, M, by hypothesis, the rectangle GY, YF will have the same ratio to the rectangle FM, MH as the given ratio a to b. Let EG be joined, meeting AB in L and the circumference again in Z; therefore since DG, EG have been inflected from the points D, E to the circumference, meeting it again in F, Z and the straight line AB in the points Y, L, again, by hypothesis, the rectangle GY, YF will have the same ratio to the rectangle GL, LZ as the ratio a to b. Therefore the rectangle GY, YF is to the rectangle FM, MH as the same rectangle GY, YF is to the rectangle GL, LZ, which is consequently equal to the rectangle FM, MH. Let the

---

[f] See the Demonstration of Proposition 63.

[*] See the last figure above.

perpendicular CK be drawn to AB, meeting the circumference in O, P, and, if K is inside the circle, let AB meet the circumference in the points A, B; therefore AK, KB are equal; and since the rectangle GL, LZ is equal to the rectangle FM, MH, the rectangle AL, LB will also be equal to the rectangle AM, MB, so that the straight line LK is equal to the straight line KM (n1).

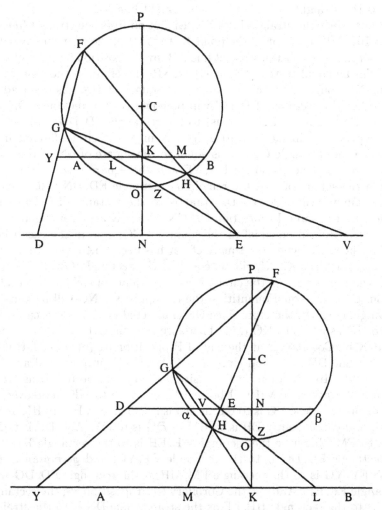

But if the point K is outside the circle, the rectangle GL, LZ is equal to the rectangle OK, KP along with the square of LK, and similarly the rectangle FM, MH is equal to the rectangle OK, KP along with the square of MK. Therefore the rectangle OK, KP along with the square of LK is equal to the rectangle OK, KP along with the square of MK; therefore the straight line LK is equal to the straight line KM (n2). It is therefore established that it is a property of the points D, E that if from one or other of them D a straight line DGF is drawn to the circle in any manner, and EG, EF are joined, these will cut off from the straight line AB, which is given in position, segments LK, MK adjacent to the point K which are equal to each other. Let DE be joined, which will be parallel to the straight line AB by Proposition 89, and let CK meet DE in N, and since CK is perpendicular to AB, then, by the Corollary to Proposition 89, the rectangle GD, DF will be to the rectangle HE, EF as the straight line DN is to the straight line NE.

Therefore since the straight lines EG, EF cut off from the straight line AB segments LK, MK adjacent to the point K, the foot of the perpendicular CK drawn from the centre to the straight line AB, by Proposition 89 the rectangle DN, NE will be equal to the rectangle ON, NP; therefore, in the case where the point N is outside the circle, when the square of DN has been added to both sides, the rectangle ED, DN will be equal (to the rectangle ON, NP along with the square of DN, that is) to the rectangle FD, DG (n3); and in the case where N is inside the circle, let DN meet the circumference in α, β, and since the rectangle ON, NP is equal to the square of αN, if the equal quantities, namely the rectangle DN, NE and the square of αN, are taken away from the square of DN, the remaining rectangle ED, DN will be equal to the remaining rectangle αD, Dβ, that is to the rectangle FD, DG (n4). And since, by the same Proposition 89, CN, CO, CK are proportional, the rectangle NC, CK will be equal to the square of CO, and, in the case where K is inside the circle, when the square of CK has been taken away from both sides, the rectangle NK, KC will be equal to the rectangle OK, KP; but in the case where K is outside the circle, when the square of CN has been taken away from the same equal quantities, the rectangle KN, NC will be equal to the rectangle ON, NP; therefore since OP is bisected in C, in both cases, PN will be to NO as PK is to KO (n5). Therefore the join GH will pass through the point K by Corollary 2 at the end of Proposition 61 (n6). Let GH meet the straight line DE in V, and since CK, CO, CN are proportional and NV is perpendicular to CN, then by Proposition 87 GV will be to VH as GK is to KH, and *permutando*, VH will be to HK as VG is to GK; moreover, on account of the parallels, VG is to GK as DG is to GY, and VH is to HK as EH is to HM; and consequently EH is to HM as DG is to GY. And DF is to FY as EF is to FM, therefore the rectangle FE, EH is to the rectangle FM, MH as the rectangle FD, DG is to the rectangle FY, YG; and *permutando*, the rectangle FY, YG is to the rectangle FM, MH as the rectangle FD, DG is to the rectangle FE, EH. And, by the Corollary to Proposition 89, the rectangle GD, DF is to the rectangle HE, EF as the straight line DN is to the straight line NE, so that also the rectangle GY, YF is to the rectangle HM, MF as the straight line DN is to the straight line NE. But by hypothesis the ratio of the rectangle GY, YF to the rectangle HM, MF is given, namely it is the same as the given ratio a to b; therefore the ratio DN to NE is given. Now since CK, CO, CN are proportional by the first part of Proposition 89 and CK, CO are given, CN will also be given, and the rectangle ON, NP, to which the rectangle DN, NE is equal [Prop. 89], will be given. Therefore since the ratio DN to NE is given, and the rectangle contained by them is given, the rectangle DN, NE is given in type and magnitude, and consequently its sides DN, NE are given in magnitude [*Dat.* 60] and also in position, and the point N is given, therefore the points D, E are given. This is what had to be shown.

It will be composed as follows.

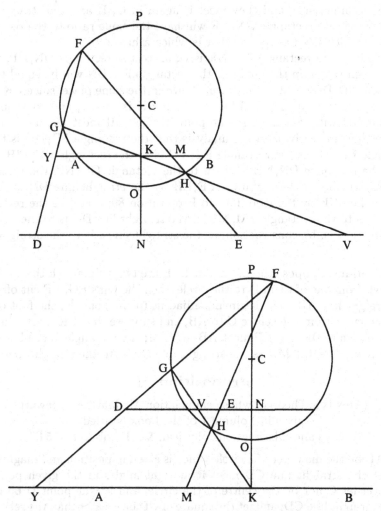

Let the straight line AB be given in position, and let the given ratio be that which the straight line a has to the straight line b. Let the perpendicular CK be drawn from the centre C to AB, and let it meet the circumference in O, P, and let a third proportional CN be taken to CK, CO, and let DE be drawn through N parallel to AB; and let a rectangle be made equal to the rectangle ON, NP and moreover similar to the rectangle which is contained by a, b; and on the straight line DN let ND be set equal to one side of the figure found, namely its side which is proportional to a; and on the other side of the point N let NE be set equal to the remaining side, if AB meets the circle, but on the same side if AB does not meet the circle. Therefore the rectangle DN, NE is equal to the rectangle ON, NP, and DN has the same ratio to NE as the straight line a has to the straight line b. Now the points D, E will be the points which have to be found, that is, if DF, EF are inflected from them to the circumference, and they meet it again in G, H and the straight line AB in Y, M, the rectangle GY, YF will have the same ratio to the rectangle HM, MF which DN has to NE, that is which a has to b.

For since the rectangle DN, NE is equal to the rectangle ON, NP, then, as has been shown in the analysis, the rectangle ED, DN will be equal to the rectangle FD, DG. And, as has been shown in the same place, since CK, CO, CN are proportional, PN will be to NO as PK is to KO, and consequently the join GH will pass through the point K. Let GH meet DE in V, and it will be shown exactly as in the analysis that the rectangle FD, DG is to the rectangle FE, EH as the rectangle FY, YG is to the rectangle FM, MH. Now since the rectangle DN, NE is equal to the rectangle ON, NP, the rectangle FD, DG will be to the rectangle FE, EH as the straight line DN is to the straight line NE by the Corollary to Proposition 89; therefore the rectangle FY, YG is to the rectangle FM, MH as the straight line DN is to the straight line NE, that is, by construction, as the straight line a is to the straight line b.

Corollary. Suppose therefore that D, E are two points such that, if DGF is drawn from one of them D to the circle, then the joins EG, EF cut off from the straight line AB equal segments adjacent to the point K, the foot of the perpendicular from the centre C to AB, and suppose that CK meets the join DE in N, while DF and EF meet AB in Y, M; the rectangle GY, YF will be to the rectangle HM, MF as the straight line DN is to the straight line NE.

(p.248)   **Proposition 91**

(This is a Theorem whose enunciation Mr Matthew Stewart
gives in Volume 1 of the book entitled
"Essays and Observations physical, &c. Edinburgh 1754.")

"Let AB be the diameter of a circle which is given in position and magnitude, and let the straight line CD meet it at right angles in the given point C; moreover let C not be the centre of the circle, and let the point D be given on the straight line CD; and let the square of CD be greater than the rectangle AC, CB, if the point C is inside the circle, but if it is outside, let the square of CD be less than the rectangle AC, CB; let DC be produced to E, so that CE is equal to CD; two points F, G not on the straight line DE will be given such that, if FH, GH are inflected from them to any point H on the circumference, meeting the straight line DE in K, L, then the sum of the squares of DK, EK will be in a given ratio to the rectangle DL, LE."

This Theorem is in fact a Porism, which I have investigated as follows.

Case 1. Here the point C is inside the circle.

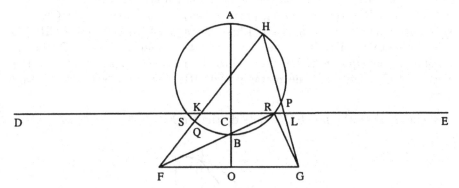

Suppose that the Porism is true, and let the straight line DE meet the circumference in the points R, S; therefore since DE is perpendicular to the diameter AB, then CR will be equal to CS; and since, by hypothesis, the ratio of the sum of the squares of DK, KE to the rectangle DL, LE is given, the ratio of half this sum, that is (n1) the ratio of the squares of KC, CE [II 9 or II 10], to the rectangle DL, LE, will also be given. Let FR, GR be inflected to the point R, and likewise the ratio of the squares of CR, CE to the rectangle DR, RE will be given. Therefore the squares of CR, CE are to the rectangle DR, RE as the squares of KC, CE are to the rectangle DL, LE; and (n2) the remaining rectangle, namely SK, KR, will be to the remaining rectangle SL, LR as the squares of CR, CE are to the rectangle DR, RE [V 19]. But DR, CR, CE are given, and consequently RE is given, therefore the ratio of the squares of CR, CE to the rectangle DR, RE is given, and consequently the ratio of the rectangle SK, KR to the rectangle SL, LR is given. Let the straight lines FH, GH meet the circumference again in Q, P, and since the rectangle SK, KR is equal to the rectangle QK, KH, and the rectangle SL, LR is equal to the rectangle PL, LH, the rectangle QK, KH will have to the rectangle PL, LH the given ratio which the sum of the squares of CR, CE has to the rectangle DR, RE. Therefore since F, G are two points such that, if FH, GH are inflected from them to the circumference in any manner, meeting it again in Q, P and the straight line DE in the points K, L, the ratio of the rectangle QK, KH to the rectangle PL, LH is the same as the given ratio which the sum of the squares of CR, CE has to the rectangle DR, RE (n3), then by Proposition 90 the points F, G will be given.

It will be composed thus.

Therefore let the points F, G be found in the way shown in Proposition 90; if FH, GH are inflected from them to the circumference in any manner, meeting the straight line DE in K, L, then the sum of the squares of DK, KE will have to the rectangle DL, LE a given ratio, namely that which twice the squares of CR, RE has to the rectangle DR, RE. For since, by construction, the rectangle QK, KH is to the rectangle PL, LH, that is the rectangle SK, KR is to the rectangle SL, LR, as the sum of the squares of CR, CE is to the rectangle DR, RE, the remaining part, namely the squares of KC, CE, will be to the remaining rectangle DL, LE as the sum of the squares of CR, CE is to the rectangle DR, RE. Therefore twice the squares of KC, CE, that is the sum of the squares of DK, KE, is to the rectangle DL, LE as twice the squares of CR, CE is to the rectangle DR, RE, that is they are in a given ratio.

Case 2. Here the point C is outside the circle.

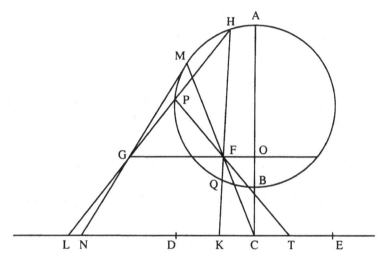

As in the first case, let straight lines FH, GH be inflected in any manner, meeting the circumference again in Q, P and the straight line DE in the points K, L, and let the join FP meet the straight line DE in T. Therefore, by hypothesis, the ratio of the sum of the squares of DK, KE to the rectangle DL, LE is given; and since the inflected straight lines FP, GP meet the straight line DE in T, L, the sum of the squares of DT, TE will be in the same given ratio to the same rectangle DL, LE. Therefore the squares of DK, KE are equal to the squares of DT, TE, and their halves, namely the squares of KC, CE and the squares of TC, CE, are equal; therefore the straight line KC is equal to the straight line CT. Therefore it is a property of the points F, G that if from one or other of them G a straight line GPH is drawn to the circle, then FH, FP, having been joined and produced, cut off from the straight line DE equal segments adjacent to the point C in which the diameter AB perpendicular to DE meets it. Let the join GF meet the straight line AB in O, and by the Corollary to Proposition 90 the rectangle PL, LH will be to the rectangle QK, KH as the straight line GO is to the straight line OF. But (n4) the rectangle PL, LH is equal to the rectangle AC, CB and the square of LC; and likewise the rectangle QK, KH is equal to the rectangle AC, CB

and the square of KC; therefore the ratio of these is the same as the ratio GO to OF. Let CF be joined and let it meet the circumference in M, and let the join MG meet the straight line DE in N. Therefore the same thing will be shown about the inflected straight lines GM, FM as has just been shown about the inflected straight lines GH, FH, namely that the rectangle AC, CB and the square of NC are to the rectangle AC, CB as the straight line GO is
* to the straight line OF. Therefore (n5) the rectangle AC, CB and the square of LC are to the rectangle AC, CB and the square of KC as the rectangle AC, CB and the square of NC are to the rectangle AC, CB; thus the remaining
† part, namely the excess of the squares of NC, LC, is to the remaining part, namely the square of KC, as the rectangle AC, CB and the square of NC are to the rectangle AC, CB. Now since the ratio of the sum of the squares of DK, KE to the rectangle DL, LE is given by hypothesis, the ratio of half of this sum, that is of the sum of the squares of KC, CE, to the same rectangle, that is to the excess of the squares of LC, CE (n6), will be given, and, by inverting, the ratio of this excess to that sum will be given. And likewise, since FM, GM have been inflected, the excess of the squares of NC, CE will be to the square of CE in the same given ratio by hypothesis. Therefore the excess of the squares of LC, CE is to the sum of the squares of KC, CE as the excess of the squares of NC, CE is to the square of CE; thus the remaining part, namely the excess of the squares of NC, LC, is to the remaining part, namely the square of KC, as the excess of the squares of NC, CE is to the square of CE. And it was shown (see †) that the excess of the squares of NC, CE is to the square of KC as the rectangle AC, CB and the square of NC are to the rectangle AC, CB; therefore the excess of the squares of NC, CE is to the square of CE as the rectangle AC, CB and the square of NC are to the rectangle AC, CB (n7). Thus, as the remaining part, namely the sum of the rectangle AC, CB and the square of CE, is to the remaining part, namely the excess of the rectangle AC, CB and the square of CE, so the rectangle AC, CB and the square of NC are to the rectangle AC, CB, and so as was shown (see *) the rectangle AC, CB and the square of LC are to the rectangle AC, CB and the square of KC, that is the rectangle PL, LH to the rectangle QK, KH (n8). But the points A, B, C, E are given; therefore the ratio of the rectangle AC, CB and the square of CE to the excess of the same quantities is given, and consequently the ratio of the rectangle PL, LH to the rectangle QK, KH is given; therefore by Proposition 90 the points F, G are given.

It will be composed thus.

Let the points F, G be found by means of Proposition 90 so that if two straight lines FH, GH are inflected from them to the circumference APB in any manner, meeting the straight line DE in the points K, L and the circumference again in Q, P, the rectangle PL, LH is to the rectangle QK, KH as the rectangle AC, CB along with the square of CE is to the excess of the rectangle AC, CB and the square of CE; then the points F, G will be the points which have to be found, that is if FH, GH have been inflected as was said, the squares of DK, KE together will be to the rectangle DL, LE in a given ratio, namely that which twice the excess of the rectangle AC, CB over the square of CE has to the rectangle AC, CB along with the square of CE.

For since, by construction, the rectangle PL, LH is to the rectangle QK, KH, that is the rectangle AC, CB along with the square of LC is to the rectangle AC, CB along with the square of KC, as the rectangle AC, CB along with the square of CE is to the excess of the rectangle AC, CB and the square of CE, the remaining part, namely the excess of the squares of LC, CE, that is the rectangle DL, LE, will be to the remaining part, namely the square of KC along with the square of CE, as the rectangle AC, CB along with the square of CE is to the excess of the same quantities; and *invertendo*, the sum of the squares of KC, CE is to the rectangle DL, LE as the excess of the rectangle AC, CB and the square of CE is to the sum of the same quantities; therefore twice the squares of KC, CE, that is the squares of DK, KE together, are to the rectangle DL, LE as twice the excess of the rectangle AC, CB and the square of CE are to the rectangle AC, CB along with the square of CE. Q.E.D.

(p.249) ## Proposition 92

"Suppose that the straight line AB has been given in position and that it meets the circle AEB which is given in position in the points A, B; two points C, D will be given such that if straight lines CE, DE are inflected from them to the circumference in any manner, meeting AB in F, G, then the square of FG will be to the rectangle AG, GB in the given ratio which the straight line a has to the straight line b. Moreover the straight line AB must not pass through the centre."

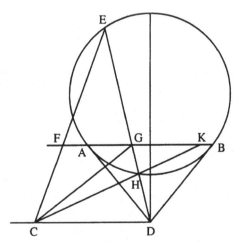

Suppose that the Porism is true; therefore, by hypothesis, the square of FG is to the rectangle AG, GB as the straight line a is to the straight line b. Let DE meet the circumference again in H, and let CH be joined, meeting

AB in K; therefore since CH, DH have been inflected from the points C, D to the circumference, again, by hypothesis, the square of GK will be to the rectangle AG, GB as a to b; therefore the square of FG is equal to the square of GK, and the straight line FG is equal to the straight line GK; that is the points C, D are such that if a straight line DHE is drawn in any manner from the point D to the circle, then the joins CE, CH intercept equal segments of the straight line AB between themselves and the straight line DE. Therefore if DA is drawn, it will necessarily be tangent to the circle at A, for the join CA takes off no segment between itself and DA, but if DA were to meet the circumference again, the straight line drawn from the point C to this point of intersection would take off some segment between itself and DA, which is not possible; thus DA is tangent to the circle; in the same way it will be shown that DB is tangent to the circle. But the points A, B are given, so that the tangents AD, BD are given in position, and the point D will be given. Now the straight line DE is cut harmonically in the points G, H [Prop. 73], therefore if CG is joined the straight lines CE, CG, CH and CD are harmonicals, and FGK is bisected by three of them, consequently the fourth CD is parallel to the straight line FK [Prop. 82]; and the point D is given, so that DC is given in position. Now since EG is to GH as ED is to DH, the square of ED will be to the rectangle ED, DH as the square of EG is to the rectangle EG, GH, that is to the rectangle AG, GB (n1), and *permutando*, the square of ED is to the square of EG as the rectangle ED, DH is to the rectangle AG, GB, and *permutando* (the square of ED is to the square of EG as the rectangle ED, DH is to the rectangle AG, GB). But the square of ED is to the square of EG as the square of CD is to that of FG (n2); thus the square of CD is to that of FG as the rectangle ED, DH, that is the square of DA, is to the rectangle AG, GB, and *permutando*, as the square of CD is to the square of DA, so the square of FG is to the rectangle AG, GB, that is, by hypothesis, in a given ratio. And the square of DA is given, therefore the square of DC is given, and the straight line DC and the point C will be given. This is what had to be shown.

It will be composed thus.

From the points A, B let AD, BD be drawn tangent to the circle, and through D let the straight line parallel to the straight line AB be drawn, and on it let DC be taken such that the square of CD is to that of DA as the straight line a is to the straight line b; then C, D will be the points which have to be found.

For let the straight lines CE, DE be inflected to the circumference in any manner, meeting the straight line AB in the points F, G. Therefore since ED is to DH as EG is to GH [Prop. 73], the square of ED will be to the rectangle ED, DH as the square of EG is to the rectangle EG, GH; and *permutando*, the rectangle ED, DH, or the square of AD, is to the rectangle EG, GH, that is to the rectangle AG, GB, as the square of ED is to the square of EG, that is as the square of CD is to that of FG; and again *permutando*, the square of

FG is to the rectangle AG, GB as the square of CD is to the square of DA, that is as the straight line a is to the straight line b. Q.E.D.

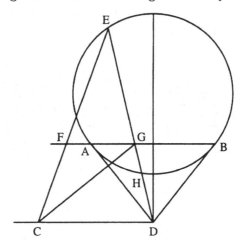

(p.249)

Proposition 93

(This and the preceding Proposition are two parts of a Porism which Mr Matthew Stewart proposed to me.)

"Suppose that the straight line AB has been given in position outside the circle EPF which is given in position, and let the straight line CD be drawn from the centre C perpendicular to AB, meeting the circumference in E, F; and on AB let the points A, B be taken equidistant from the point D, such that the square of AD, or of DB, is equal to the rectangle ED, DF. Two points G, H will be given such that if straight lines GK, KH are inflected from them in any manner to the circumference, meeting the straight line AB in the points L, M, then the square of LM will be to the squares of MA, MB together in the given ratio which the straight line a has to twice the straight line b."

Suppose that the Porism is true; therefore, by hypothesis, the square of LM is to the squares of AM, MB together as the straight line a is to twice the straight line b, so that also the square of LM is to half the squares of AM, MB, that is (n1) to the squares of AD, DM together [II 9 or II 10], as the straight line a is to the straight line b. Let HK meet the circumference again in N, and let the join GN meet the straight line AB in O; therefore since GN, HN have been inflected from the points G, H, meeting the straight line AB in O, M, it will be shown similarly that the square of OM is to the squares of AD, DM together as the straight line a is to the straight line b. Therefore the square of LM is equal to the square of OM, and the straight line LM is equal to the straight line OM. It is therefore a property of the points G, H that if a straight line HK is drawn through H in any manner, meeting the circumference in K, N, and if GK, GN are joined and produced, they cut off equal segments LM,

OM of the straight line AB between themselves and the straight line KHM; and it follows from this that the join GH is parallel to the straight line AB; for if this is not so, let GR be drawn parallel to AB, and let it meet the circumference in R, and let the join RH meet the circumference again in S, and let GS be joined; therefore since RS has been drawn through the point H, the straight lines GR, GS will cut off equal segments of the straight line AB between themselves and the straight line RHS, and certainly the straight line GS cuts off some segment between itself and the straight line RS, since both meet the straight line AB; but GR, which is parallel to AB, can cut off no segment between itself and the straight line RS, which is impossible;

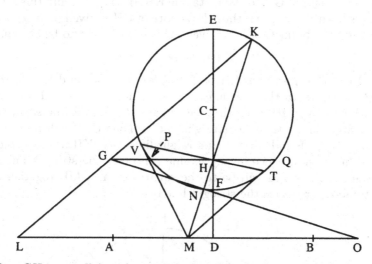

therefore GH is parallel to the straight line AB. And since LM, MO are equal, and AB, GH are parallel, as KM is to KH so (LM will be to GH, that is MO will be to GH, and so) MN will be to NH; therefore the straight line KM is cut harmonically in the points N, H. From the point M let straight lines MT, MV be drawn tangent to the circle, and let TV be joined; therefore the straight line MK is cut harmonically in the point N and the point in which it meets the straight line VT [Prop. 73]. Therefore the point H is on the straight line VT. Likewise, if two straight lines are inflected from the points G, H to another point on the circumference, and if the one which is inflected from the point H meets the straight line AB in X, it will be shown that the point H is on the straight line which joins the points of contact of the tangents which are drawn from the point X; therefore the point H is at the intersection of this straight line and the straight line VT. Now the straight line VT, as also any other straight line which joins the points of contact of the tangents which are drawn from any point in AB, passes through the point on the diameter CD whose distance from C is a third proportional to the straight lines CD, CF, as is shown in the Lemma at the end of this Proposition. Therefore the point H is on the straight line CD, and CD, CF, CH are proportional. But

CD, CF are given, consequently CH is given in magnitude, but it is also given in position; thus the point H is given, as also is the rectangle KH, HN (n2). But since MK is to KH as MN is to NH, the rectangle KM, MN will be to the rectangle KH, HN as the square of MK is to the square of KH, that is as the square of LM is to that of GH; and *permutando*, the square of GH will be to the rectangle KH, HN as the square of LM is to the rectangle KM, MN. Now the rectangle KM, MN is equal to the rectangle ED, DF along with the square of DM, that is to the squares of AD, DM (n3); and, by hypothesis, the ratio of the square of LM to the squares of AD, DM is given, namely it is the same as the ratio of the straight line a to the straight line b. Therefore the ratio of the square of GH to the rectangle KH, HN is given and this rectangle has been shown to be given; thus the square of GH is given, and the straight line GH and the point G will be given. This is what had to be shown.

It will be composed thus.

Let a third proportional CH be found for CD, CF, and let the straight line HP be drawn to the circumference at right angles to CH, and let the point G be taken in HP such that the square of GH is to the square of HP as the straight line a is to the straight line b. Then G, H will be the points which have to be found; that is, if straight lines GK, KH are inflected from them to the circumference in any manner, meeting the straight line AB in L, M, the square of LM will be to the squares of AM, MB together as the straight line a is to twice the straight line b.

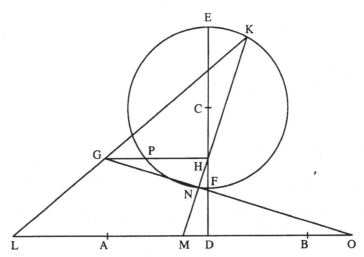

For since CD, CF, CH are proportional, and KHM has been drawn through H to the straight line AB which is perpendicular to CD, then (n4) MK will be to KH as MN is to NH [Prop. 87]; thus as the square of MK is to the square of KH, that is as the square of LM is to that of GH, so [VI 22] the rectangle KM, MN, that is the rectangle ED, DF, or the square of AD, along with the square of DM is to the rectangle KH, HN, that is to the square of PH; and *permutando*, as the square of LM is to the squares of AD, DM together so the square of GH is to that of PH, and so, by construction, is the straight line a to the straight line b. Therefore the square of LM is to twice the squares of AD, DM, that is to the squares of AM, MB together, as the straight line a is to twice the straight line b. Q.E.D.

Lemma to Proposition 93

"Let the straight line AB be outside the circle whose centre is C, and let CD be drawn perpendicular to AB; if straight lines are drawn from any point E on AB, tangent to the circle in the points F, G, and the join FG meets the straight line CD in H, while CD meets the circumference at the point K between C and D, then CD, CK, CH will be proportional."

For let EC be joined and let it meet the straight line FG in L. Therefore since the angles ELH, EDH are right angles, the points E, L, H, D are on a circle, so that the rectangle DC, CH is equal to the rectangle EC, CL, that is to the square of the semidiameter CK; therefore CD, CK, CH are proportional.

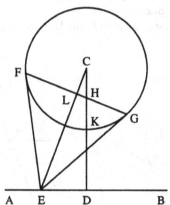

And if a straight line FHG is drawn in any manner through a point H inside the circle, and from the points F, G where it meets the circumference straight lines FE, GE are drawn tangent to the circle, and through their point of intersection E the straight line ED is drawn perpendicular to the straight line CH produced, then likewise CD, CK, CH will be proportional; this is demonstrated in the same way as the preceding assertion.

Notes on Part IV

Note on Simson's introductory comments (p. 201). Simson has already commented on Fermat's Propositions in his Preface (pp. 13–14) and his introductory material (pp. 31–32). See also my note on Simson's Preface, p. 38. Concerning the "ratio of the straight line AZ to the fourth part of ZD" see my note on Proposition 85 (p. 245) (composition and Fermat's statement of his fifth Proposition).

Note on Proposition 80 (p. 201). (n1) In the final part of the composition from $\frac{KA}{HL} = \frac{AG}{AH}$ we get $\frac{KA}{HL} = \frac{KA+AG}{HL+AH} = \frac{KG}{AL}$ as claimed.

(n2) Simson's final assertion[24] that DE should not be parallel to AC is curious. Certainly in his analysis he uses the point of intersection O of AC and DE to deduce that KP.PL = KA.AL. However, if AC and DE are parallel, O is a point at infinity and the same identity can be deduced as a limit. Such a process would probably be objectionable to Simson (but see the note on Proposition 28, p. 188). His composition does not depend on the assumption that AC and DE are not parallel.

An analysis for the case where AC and DE are parallel might be given as follows.

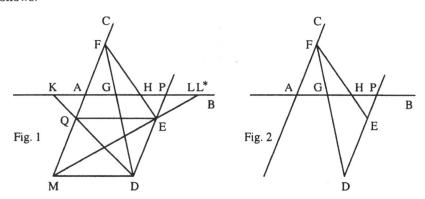

Fig. 1

Fig. 2

Following Simson we obtain KG.HL = KA.AL, so we just have to show that K and L are given. Suppose first that K does not coincide with P (Fig. 1). Join KD and let it meet AC in Q. Join EQ. Then EQ must be parallel to AB, since otherwise QE would cut off a finite segment adjacent to L, while DQ cuts off a zero segment adjacent to K, so that the required product is zero. Since E is given, Q must then be given and consequently K is given. Now through D draw a line parallel to AB and meeting AC in M. Join ME (which cannot be parallel to AB) and let it meet AB in L*. Then KG.HL is fixed by hypothesis and, from Simson's composition, KG.HL* is also fixed. Therefore

---

[24] "Notandum autem est rectam DE non debere parallelam esse rectae AC."

the two quantities KG.(HL ± HL*) are fixed. But one of these is KG.LL*, which varies with G unless L and L* coincide. Thus L is given, since M, E and therefore L* are given. We show similarly that if L does not coincide with P then again K and L are given.

Suppose then that both K and L coincide with P (Fig. 2), so that GP.HP is fixed and equal to AP^2. This is clearly impossible.

Fermat's enunciation is as follows (translation):

"Let there be two straight lines ON, OC, which form an angle at the point O and let them be given in position, let also points A and B be given, from the points B and A let straight lines BE, AF be drawn parallel to OC and meeting the straight line NO produced in the points E and F, let the straight line AE be joined, which meets the straight line CO produced in D, likewise let the straight line FB be joined, which meets the same straight line CO in C and to any point of the straight line ON, say V, let straight lines AV, BV be inflected, such that the straight line AV meets the straight line OC in the point S, while the straight line BV meets the same OC in the point R; then the rectangle CR, DS will always be equal to the rectangle CO, OD, and so will be equal to a given area."

Note on Proposition 81 (p. 203). (n1) In the first paragraph of the analysis from $\frac{CF.ED}{CE.FD} = \frac{HG}{GK}$ we get $\frac{CF.ED-CE.FD}{CE.FD} = \frac{HG-GK}{GK} = \frac{HK}{GK}$ and CF.ED − CE.FD = CF.(CD − CE) − CE.(CD − CF) = CF.CD − CE.CD = EF.CD, so that $\frac{EF.CD}{CE.FD} = \frac{HK}{GK}$ as claimed.

(n2) Note that CD is not required to be a diameter of the circle, although in the composition and subsequent discussion Simson twice refers to it in this way and his original diagram does appear to show CD as a diameter. This aberration is no doubt carried over from Fermat, who did require the line to be a diameter; his formulation of the proposition is as follows (translation):[25]

"Let there be a circle whose diameter is the straight line AD and let NM be drawn parallel to AD in any manner to meet the circle in the points N and M; and let the points N and M be given, let straight lines NB, BM be inflected in any manner and let them cut the diameter in the points O and V. I say that the ratio of the rectangle AO, DV to the rectangle AV, DO is given; and so if NC, CM are inflected cutting the diameter in the points R, S, the rectangle AR, DS will always be to the rectangle AS, DR as the rectangle AN, DV is to the rectangle AV, DO, and it is not difficult to extend the proposition to the ellipse, the hyperbola and opposite sections."

Note on Proposition 83. (p. 205) Case 1. (n1) Note that the result is false if we take the other pair of segments which might be inferred from the enunciation, viz. AF, GB. Since $AF^2 + GB^2 = (AG - FG)^2 + (FB - FG)^2 =$

---

[25]The notation has been changed slightly to accord with that used by Simson: Fermat refers for example to "recta NBM" by which he means two straight lines NB and BM.

$AG^2 + FB^2 - 2(AG + FB - FG).FG = AG^2 + FB^2 - 2AB.FG$ and Simson shows that $AG^2 + FB^2$ is given, then $AF^2 + GB^2$ must vary with E.

(n2) Simson's assertion that DE, DM, DH, DC are harmonicals means in modern terminology that they form a harmonic pencil.

The net result of Case 1 is that AC is perpendicular to AB and AC $= AB/\sqrt{2}$.

(n3) The paper of Euler's to which Simson refers is 'Variae demonstrationes geometriae' [6] (see also the summary[26] on the introductory pages 37–38 of the journal volume), which is also found in Volume 26, pp. 15–32, of [7]. Euler's discussion of Case 1 of Proposition 83 occurs on pp. 49–52 of the journal volume and is a direct proof which is not based on an initial analysis. He noted that the proposition had been proposed for proof by Fermat in correspondence and claimed that a proof had not previously been published.

Case 2. Again as in Case 1 the segments must be chosen appropriately: they must be DL and EK, not DK and EL.

(n4) That $FC.CQ = (radius)^2$ comes from the fact that PO is the polar of F with respect to the given circle.

(n5) The point R on AB is determined from the condition $\frac{CD}{CA} = \frac{CA}{CR}$ and then RF is perpendicular to AR with $RF = RZ = \sqrt{(CR^2 + CZ^2)} = \sqrt{(CR^2 + (AB/2)^2)}$. If D coincides with A as in Case 1 we must then have that R and A coincide and $RF = \sqrt{((AB/2)^2 + (AB/2)^2)} = AB/\sqrt{2}$, as in Case 1.

(n6) Simson asserts that $Hd^2 + df^2 = dg^2$ and $eH^2 + Hf^2 = 2\,dg^2$; since $Hd = dc$, $ed = df$ and it has been shown that $df^2 = Hg.gc$, we deduce $Hd^2 + df^2 = Hd^2 + (dg - Hd).(dg + Hd) = dg^2$ and $eH^2 + Hf^2 = (df - Hd)^2 + (Hd^2 + df^2)^2 = 2(Hd^2 + df^2) = 2\,dg^2$.

(n7) Note that in the Corollary, while $kn^2 + ml^2$ is independent of H, it does depend on the particular line drawn parallel to AB.

Fermat's enunciation is as follows (translation):

"Let the circle ICH be displayed whose diameter IDH is given, centre is D, radius normal to the diameter is CD, let points B and A be taken as given in the diameter produced, and let the straight lines AI, BH be equal, let DL to LI be made as DI to IA, and let the straight line DR be equal to DL; the points R and L will be given; let the straight line CA be joined and let AF be set equal to this and perpendicular to the diameter, and let BG be made equal and parallel to AF, let any straight lines FE, EG be inflected to the circle from the points F and G, cutting the diameter in the points M and N; I say that the sum of the two squares of RM, LN is always equal to the same given area; the same things having been put in place, in the second case let the straight line CL be joined and let LP be set equal to this and perpendicular to the diameter, and let RZ be made equal and parallel to LP; if from the two points Z and P any straight lines PV, VZ are inflected to the

---

[26]Here and in the list of papers the word "geometriae" in the title has been replaced by "geometricae".

circumference cutting the diameter in the points K and T, the aggregate of the squares of AT and BK will always be equal to another given area."

Note on Proposition 84 (p. 213). (n1) In Case 2 of Proposition 83 Simson discusses the proportionality of CD, CA, CR and CV, CY, CX (see p. 209). Since CR and GX are equal, this is presumably what he is referring to in the second paragraph of his discussion of Proposition 84.

(n2) In the third paragraph the assertion that $Va^2 = XV.VC$ comes from the fact that triangles XVa and aVC are similar; (n3) that $\frac{CV^2}{aV.VC} = \frac{kh.AC}{kh.XG}$ is immediate from the established ratio $\frac{AC}{XG} = \frac{CV}{Va}$ on multiplying top and bottom by kh on the left and CV on the right.

Note that in the composition, although F is introduced at the beginning, it is not specified until the final paragraph.

(n4) In the penultimate paragraph Simson's reference to similar rectangles is equivalent to the following: since triangles DCV and GXV are similar, $\frac{XG}{XV} = \frac{DC}{VC}$ and so $\frac{XG.DC}{XV.VC} = \frac{DC^2}{VC^2} = \frac{AC^2}{aV^2}$.

Note on Proposition 85 (p. 218). The main points of Simson's analysis which require comment are the following.

(n1) From $DF^2 = DK^2 + KF^2$ and $FE^2 = FK^2 + KE^2$ we get $DF^2 + FE^2 = DK^2 + KE^2 + 2FK^2$. Then using $DK^2 = (DH - HK)^2 = DH^2 + HK^2 - 2DH.HK$ and $KE^2 = (KH + HE)^2 = (KH + DH)^2 = KH^2 + DH^2 + 2KH.DH$ we deduce $DF^2 + FE^2 = 2(DH^2 + HK^2 + KF^2) = 2(DH^2 + HF^2)$. Similarly $DG^2 + GE^2 = 2(DH^2 + HG^2)$ and consequently $HF = HG$.

(n2) Then since the area of triangle DFE is $\frac{1}{2}DE.FK$ and by hypothesis $\frac{2(DH^2 + HF^2)}{\text{area of } \triangle DFE} = \frac{2HN}{HD}$, we get $\frac{DH^2 + HF^2}{DE.FK} = \frac{HN}{2HD} = \frac{HN}{DE}$ and so $DH^2 + HF^2 = HN.FK = NH.HM$. This yields $DH^2 + HF^2 - HM^2 = NH.HM - HM^2$, which reduces to $DH^2 + FM^2 = (NH - HM).HM = HM.NM$.

(n3) If F is moved to O, the point M will then coincide with O and the last relation will become $DH^2 = HO.NO$, while if F moves to P, the point M will coincide with P and we get $DH^2 = HP.NP$. Consequently $HO.NO = HP.NP$ and so, since the points are collinear, the two smaller segments must be equal and the two larger segments must be equal, i.e. $HO = NP$ and $NO = HP$.

(n4) In the final paragraph of the analysis from $DH^2 + FM^2 + MC^2 = HM.MN + MC^2$ we get $DH^2 + FC^2 = (HC - CM).(NC + CM) + MC^2 = (HC - CM).(HC + CM) + MC^2 = HC^2$.

The composition uses similar ideas. The final assertion comes from $\frac{4RT}{RC} = \frac{4QP}{QC} = \frac{4CH}{DH} = \frac{2NH}{DH}$.

The Corollary depends on the fact that triangles XFV and DFE are similar.

Fermat's statement of his fifth Proposition is as follows (translation):

"Let there be a circle RAC, whose diameter RDC is given, centre is D, radius DA is normal to the diameter, and let given points Z and B be taken arbitrarily in the diameter equidistant from the centre D, and, AZ having been joined, let ZM be made equal to AZ and perpendicular to the diameter, and let BO be drawn equal and parallel to ZM, let any straight lines MH, OH be inflected to the circumference cutting the diameter in the points E and N; the ratio of the squares of EH, HC taken together to the triangle EHN will always be given, namely it will be the same as that of the straight line AZ to the fourth part of the straight line ZD."

Note on Proposition 87 (p. 221). The points K, N are inverse points with respect to the circle and the conclusion is that $\{QR, KE\}$ is a harmonic set. In Case 1 Simson deduces that E and I are also inverses of each other, so that KI lies on the polar of E.

(n) In the last paragraph of the demonstration Simson argues from $SK.KE = CK.KN$ and $CS^2 + SK^2 = CK^2$ that, if E is outside the circle, then $CS^2 + SK^2 + SK.KE = CK^2 + CK.KN$, therefore $CS^2 + SK.(SK + KE) = CK.(CK + KN)$, and so $CS^2 + SK.SE = CK.CN$, while if E is inside the circle, the same conclusion follows in a similar fashion from $CS^2 + SK^2 - SK.KE = CK^2 - CK.KN$; then in both cases, since it has been shown that $ES.SK = SR^2$, we obtain $CK.CN = CS^2 + SK.SE = CS^2 + SR^2 = CR^2 = CO^2$.

In Volume Q of the *Adversaria* [43], pp. 80–82 and 47 Simson establishes the converse of Proposition 87. The articles are dated 17, 23 June and 20 September 1764. This is the only instance where I have found a direct reference[27] in the *Adversaria* to a result in the *Treatise*.

Note on Proposition 88 (p. 223). (n) If N is outside the circle we have $CN^2 - NC.CK = CN^2 - CO^2$, so that $CN.(CN - CK) = (CN - CO).(CN + CO) = (CN - CO).(CN + CP)$ and therefore $CN.NK = ON.NP$. If N is inside the circle, the same conclusion is reached by taking the differences in the other order.

Note on Proposition 89 (p. 224).

Part 1. (n1) Having shown that DQ is a tangent to the circle we should note that we can show in the same way that DR is also a tangent to the circle.

(n2) The equality of angles CDN and NKE comes in the first figure from the fact that angle CDN is an angle in the cyclic quadrilateral KSDN and angle NKE is supplementary to the opposite angle SKN, while in the second figure they are angles subtended at the circumference by the same chord SN.

(n3) The assertion that $DC.CS = CO^2$ follows because D and S are inverse points with respect to the circle.

Simson's reference to Proposition 88 should really be to its demonstration.

Part 2. Obviously in the first figure it does not matter which of the points D, E is chosen for the point through which the straight line is to be drawn. In

---

[27] "Haec conversa est Prop. 87 Porismatum."

the second figure D is outside the circle and E is inside it; Simson discusses the case where E is chosen in the final paragraph.

Note that in both figures DX will pass through the point W where GL meets the circle again and GH, FW will pass through K; these observations follow from the harmonic properties of complete quadrilaterals.

(n4) In the case of the Corollary where DE meets the circle Simson uses the following identities: $GD.DF = \alpha D.D\beta = (DN - \alpha N).(DN + \alpha N) = DN^2 - \alpha N.N\beta = DN^2 - ON.NP = DN^2 - DN.NE = DN.(DN - NE) = DN.DE$ and $HE.EF = \alpha E.E\beta = (\alpha N - EN).(\alpha N + EN) = \alpha N^2 - EN^2 = ON.NP - EN^2 = DN.EN - EN^2 = (DN - EN).EN = DE.EN$.

Note on Proposition 90 (p. 228). The straight lines a, b which are mentioned in the enunciation have no significance other than to establish the ratio a : b. In the original of the second diagram a, b have also been used to denote the points where DE meets the circle; here they have been replaced by α, β in this usage to avoid confusion. The points A, B serve only to determine the position of the straight line AB; however, in the case where the line meets the circle, Simson finds it convenient to take the points of intersection as A, B.

(n1) Simson first shows that $GL.LZ = FM.MH$. Then for K inside the circle he argues that $AL.LB = GL.LZ = FM.MH = AM.MB$ and so, since $AL + LB = AM + MB = AB$, it follows that $AL = MB$ and LM is bisected at K, thus $LK = KM$.

(n2) For K outside the circle Simson obtains this equality from $GL.LZ = OK.KP + LK^2$ and $FM.MH = OK.KP + MK^2$. For these identities see the note on Proposition 59, pp. 194–195.

(n3) In the second paragraph of the demonstration Simson obtains $DN.NE = ON.NP$ from Proposition 89. Then for N outside the circle $DN.NE + DN^2 = DN.(NE + DN) = DN.DE$ and $ON.NP + DN^2 = FD.DG$ (see the note on Proposition 59), so that $DN.DE = FD.DG$.

(n4) For N inside the circle this is obtained from $DN.NE = ON.NP = \alpha N.N\beta = \alpha N^2$, $DN^2 - DN.NE = DN.(DN - NE) = DN.DE$ and $DN^2 - \alpha N^2 = (DN - \alpha N).(DN + N\beta) = D\alpha.D\beta = DG.DF$.

(n5) Simson then deduces in a similar way from $NC.CK = CO^2$ that $NK.KC = OK.KP$ if K is inside the circle, while $KN.NC = ON.NP$ if K is outside the circle. As a result of this and the fact that OP is bisected at C he asserts that $\frac{PN}{NO} = \frac{PK}{KO}$, in other words {PO, NK} is a harmonic set. Again he is probably applying what for him would have been a standard result. For K inside the circle we have $\frac{PN}{NO} = \frac{PK}{KO} \iff \frac{PK+KN}{KN-KO} = \frac{PK}{KO} \iff PK.KO + KN.KO = PK.KN - PK.KO \iff 2PK.KO = KN.(PK - KO) = 2KN.CK$. For K outside the circle the corresponding argument is: $\frac{PN}{NO} = \frac{PK}{KO} \iff \frac{PN}{NO} = \frac{PN+NK}{NK-NO} \iff PN.NK - PN.NO = NO.PN + NO.NK \iff 2PN.NO = NK.(PN - NO) = 2NK.CN$.

(n6) This harmonic property together with the equality DN.DE = FD.DG established previously is what is needed for the application of Corollary 2 of Proposition 61 to deduce that GH passes through K.

The rest of the analysis is quite straightforward. The points D, E have been characterised as the points on the line through N perpendicular to CO such that DN.NE = ON.NP and DN : NE = a : b. This is the starting point of the composition.

Note on Proposition 91 (p. 232). Stewart stated this result without proof as the final item in his paper 'Pappi Alexandrini collectionum mathematicarum libri quarti propositio quarta generalior facta, cui propositiones aliquot eodem spectantes adjiciuntur' [37]. Although he gave no proof, Stewart regarded the result as being of some consequence (". . . observatione quidem haud indignum;") and went on to say that there is a version for conic sections and that other new results may be derived from it. This publication is also known as the *Edinburgh Physical Essays* and recorded the proceedings of the Philosophical Society of Edinburgh, which MacLaurin had been largely instrumental in forming in 1737 and which subsequently developed into the Royal Society of Edinburgh. It is probable that Stewart's paper was read to the Society much earlier than its publication date.

Case 1. (n1) That $\frac{1}{2}(DK^2+KE^2) = KC^2+CE^2$ comes from $DK^2+KE^2 = (DC - KC)^2 + (KC + CE)^2 = (CE - KC)^2 + (KC + CE)^2 = 2(KC^2 + CE^2)$.

(n2) Now $CR^2 - KC^2 = (CR - KC).(CR + KC) = (SC - KC).(CR + KC) = SK.KR$ and $DR.RE - DL.LE = (DL - RL).(RL + LE) - DL.LE = DL.RL - RL^2 - RL.LE = RL.(DL - RL - LE) = RL.(DL - RE) = RL.(DL - DS) = RL.SL$. Then from $\frac{CR^2+CE^2}{DR.RE} = \frac{KC^2+CE^2}{DL.LE}$ we obtain $\frac{CR^2+CE^2}{DR.RE} = \frac{(CR^2+CE^2)-(KC^2+CE^2)}{DR.RE-DL.LE} = \frac{SK.KR}{RL.SL}$.

(n3) The given ratio a : b which is required for the application of Proposition 90 is $r = \frac{CR^2+CE^2}{DR.RE} = \frac{CR^2+DC^2}{DC^2-CR^2}$, giving $DC^2 = CE^2 = \frac{r+1}{r-1}CR^2$. Note also that from the demonstration of Proposition 90 we have $r = \frac{FO}{OG}$ (this corresponds to $\frac{DN}{NE}$ in Proposition 90).

The composition consists of reversing some of the above steps.

Case 2. (n4) For the equalities PL.LH = AC.CB + LC² and QK.KH = AC.CB + KC² see the note on Proposition 59, pp. 194–195.

Note that NM must be a tangent to the circle. Since $\frac{PL.LH}{QK.KH} = \frac{AC.CB+LC^2}{AC.CB+KC^2}$ is constant ($= \frac{GO}{OF}$) for all positions of H and AC.CB + KC² takes on its minimum value AC.CB when H coincides with M, then AC.CB + LC² and therefore also LC must have their minimum values when H and M coincide; clearly LC is least when LG is tangent to the circle.

(n5) Equating the general case and the case where H and M coincide, we obtain Simson's item (∗): $\frac{AC.CB+LC^2}{AC.CB+KC^2} = \frac{AC.CB+NC^2}{AC.CB}$. Then $$\frac{(AC.CB+LC^2)-(AC.CB+NC^2)}{(AC.CB+KC^2)-AC.CB} = \frac{AC.CB+NC^2}{AC.CB},$$

so that $\frac{LC^2-NC^2}{KC^2} = \frac{AC.CB+NC^2}{AC.CB}$, which is Simson's item (†).

(n6) As in Case 1 we show that $\frac{1}{2}(DK^2 + KE^2) = KC^2 + CE^2$ and also $DL.LE = (LC - DC).(LC + CE) = (LC - CE).(LC + CE) = LC^2 - CE^2$.

(n7) In the case where H coincides with M the right-hand sides become CE^2 and $NC^2 - CE^2$ respectively; thus $\frac{NC^2-CE^2}{CE^2} = \frac{LC^2-CE^2}{KC^2+CE^2}$ and consequently
$$\frac{(LC^2-CE^2)-(NC^2-CE^2)}{(KC^2+CE^2)-CE^2} = \frac{NC^2-CE^2}{CE^2},$$
so that $\frac{NC^2-CE^2}{CE^2} = \frac{LC^2-NC^2}{KC^2} = \frac{AC.CB+NC^2}{AC.CB}$ by (†).

(n8) Repeating the operation we get
$$\frac{(AC.CB+NC^2)-(NC^2-CE^2)}{AC.CB-CE^2} = \frac{AC.CB+NC^2}{AC.CB},$$
so that, using (∗), we have $\frac{AC.CB+CE^2}{AC.CB-CE^2} = \frac{AC.CB+NC^2}{AC.CB} = \frac{AC.CB+LC^2}{AC.CB+KC^2} = \frac{PL.LH}{QK.KH}$.

This time the given ratio a : b which is required for the application of Proposition 90 is $r = \frac{AC.CB+CE^2}{AC.CB-CE^2} = \frac{GO}{OF}$, giving $DC^2 = CE^2 = \frac{r-1}{r+1} AC.CB$.

Again the composition is a routine reversal of some of the above steps.

Note on Proposition 92 (p. 236). Again the straight lines a, b serve only to determine the given ratio a : b. The proof is straightforward and we note only that (n1) EG.GH = AG.GB because the chords AB, EH intersect at G and (n2) $\frac{ED}{EG} = \frac{CD}{FG}$ because triangles ECD and EFG are similar, FG being parallel to CD. In one place Simson ends a sentence with *permutando*, presumably leaving it to the reader to determine the result of this operation; I have supplied it within angled brackets at the relevant place in the demonstration.

Note on Proposition 93 (p. 238). Again a, b are only required to determine the given ratio.

(n1) We have $\frac{1}{2}(AM^2 + MB^2) = \frac{1}{2}\left((AD - MD)^2 + (AD + MD)^2\right) = AD^2 + MD^2$.

(n2) From intersecting chords KH.HN = PH.HQ, which is given.

(n3) From $\frac{MK}{KH} = \frac{MN}{NH}$ (harmonic property) and $\frac{LM}{GH} = \frac{KH}{KM}$ (similar triangles) we obtain $\frac{LM^2}{GH^2} = \frac{MK^2}{KH^2} = \frac{MK.MN}{KH.NH}$, from which it follows that $\frac{GH^2}{KH.NH} = \frac{LM^2}{MK.MN} = \frac{LM^2}{ED.DF+DM^2} = \frac{LM^2}{AD^2+DM^2}$, where in the last two steps we have used the property discussed in the note on Proposition 59, pp. 194–195 and the hypothesis.

(n4) In the composition, application of Proposition 87 gives $\frac{KM}{MN} = \frac{KH}{HN}$, so that *permutando*, $\frac{KM}{KH} = \frac{MN}{HN}$, as required.

Appendices

A1. A Translation of Simson's 1723 Paper along with some Comments

The main part of the paper [26] contains Simson's restorations of the two general Propositions which Pappus mentions when discussing Euclid's Porisms in the preface to Book 7 (see pp. 33–37). According to Pappus the first Proposition has ten cases or loci, which Simson considers in Propositions 7, 8, 10–16 and 19 of the *Treatise*. Proposition 19 contains the tenth locus which Simson describes as "the most general of all"; this is the case considered in the paper. Simson gives two demonstrations of it in the *Treatise*, the second of which is a version of the demonstration given in the paper. The second Proposition is Proposition 21 of the *Treatise*, where it is presented in essentially the same way.

Next in the paper is a proposed restoration of the first Porism from Book 1 of Euclid's Porisms – this is the only one for which we have other than a fragmentary statement from Pappus. In the *Treatise*, however, Simson concedes that he had misunderstood this Porism when he gave the version in the paper and presents a revised version in Proposition 23. The two interpretations are compared below. Pappus's first Lemma (Proposition 127 in Book 7), which is applied in the paper, is Proposition 22 in the *Treatise*.

Finally, Simson considers the second Porism, for which Pappus records only a conclusion and says nothing of the circumstances in which it is to hold. It is curious that Simson makes no explicit reference to this Porism in the *Treatise*, although he does discuss in detail Pappus's second Lemma (Proposition 128 in Book 7), which he applies in his treatment of the second Porism in the paper (see Proposition 8, Article 2 of Case 4).

Some of the events leading up to the publication of Simson's paper are described in my Introduction. In the original all the diagrams are found on a separate insert. Here I have included them at appropriate places in the text.

Simson's 1723 Paper

Two general Propositions of Pappus of Alexandria, *by which he has com-prehended many of* Euclid's *Porisms, restored by the most learned* Robert Simson, *Professor of Mathematics at* Glasgow. *See* Pappus's *preface to Book 7 of the Mathematical Collections which is presented by the most distinguished* Halley *at pages 8 & 34 of* Apollonius's *Two Books on the Cutting off of a Ratio.*

$$\text{'}E\grave{\alpha}\nu \; \acute{\upsilon}\pi\tau\acute{\iota}o\upsilon \; \mathring{\eta} \; \pi\alpha\rho\upsilon\pi\tau\acute{\iota}o\upsilon \; \mathring{\eta} \; \pi\alpha\rho\alpha\lambda\lambda\acute{\eta}\lambda o\upsilon, \; etc.$$

Pappus's text, which has been truncated by the injury of time in this and the following Proposition, seems to require to be restored in the following manner.

"[Suppose that two straight lines are drawn to two straight lines, which meet each other or are parallel to each other,] and three points [a] are given on one of them [or two points, if the straight line on which they lie is parallel to one of the three remaining lines]: and let all but one of the remaining points lie on a straight line which is given in position; [b] then this point will also lie on a straight line which is given in position."

Now this is stated for only four straight lines, of which no more than two pass through the same point. But a Proposition of this type is not known in the case of an arbitrary number of straight lines, although it may be true.

"Suppose that any number of straight lines meet each other, and that there are no more than two through the same point; and let all the points on one of them be given, and let each individual point on another lie on a straight line which is given in position;" or more generally thus.

Suppose that any number of straight lines meet each other, and that there are no more than two through the same point, and further all the points on one of them are given; the number of points left over will be a triangular number, whose side shows the number of points which lie on a straight line which is given in position; if no three of these points of intersection are at the angles of a triangular area, [no four are at the angles of a quadrilateral, no five, etc. that is to say in general, none of these points of intersection form an orbit], then each remaining point of intersection will lie on a straight line which is given in position.

According to *Pappus* the first Proposition is divided into ten cases of which we shall present the demonstration of that one where none of the four lines are parallel, and the straight lines which are given in position do not pass through given points; for this case is the most general of all, and its demonstration is altogether necessary for the demonstration of the second Proposition.

---

[a] That is to say, three points of intersection.

[b] That is, one point lies on some straight line which is given in position, and another point lies on another straight line which is given in position, etc.

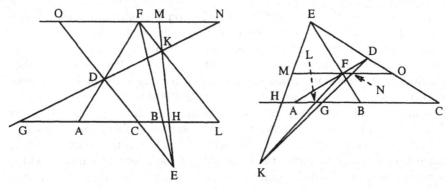

Therefore let there be four straight lines AB, AD, BE, CE. And let three points of intersection A, B, C be given on any one of them, while of the remaining points of intersection D, E, F, one D lies on the straight line GK which is given in position, another E lies on the straight line HK which is given in position; the remaining point F will also lie on a straight line which is given in position. Through F let the straight line MF be drawn parallel to AB, and let it meet HK, KG, CE in M, N, O. Therefore since the ratio HB to BC is given, the ratio MF to FO, which is equal to it, will be given, and since the ratio AC to AG is given, the ratio FO to FN, which is equal to it, will be given; hence the ratio MF to FN is given, therefore if FK is joined and meets AB in L, the ratio HL to LG will be given; and HG is given in position and magnitude; hence the point L is given, and the point K is given, therefore KL is given in position.

Therefore let HL be to LG in the ratio compounded from the ratios HB to BC and AC to AG, and let KL be joined; this will be the straight line on which the point F lies, that is to say, if any straight line CE is drawn, meeting the straight lines which are given in position in D, E, and AD, BE are joined and meet each other in F, there will be a straight line which passes through K, F, L. For let MF be drawn through F parallel to AB; since the ratio MF to FN is compounded from the ratios MF to FO and FO to FN, that is, the ratios HB to BC and AC to AG, from which the ratio HL to LG is also compounded, HL will be to LG as MF is to FN, and therefore HG will be to MN, that is HK will be to MK, as HL is to MF: hence there is a straight line which passes through K, F, L by [I 14] or [VI 32] of the Elements.

Explanation of the Second Proposition

Here it is to be noted that the number of points of intersection, which are found on one straight line in any proposed collection of straight lines, no more than two of which pass through the same point, and none of which are parallel, is one less than the number of straight lines: for two cut one another in a unique point, then the third line drawn cuts the first two in two points, the fourth cuts the first three in three points, etc. And therefore the number of points of intersection on three straight lines is one increased by two, i.e. three; the corresponding number on four straight lines is three increased by three; then on five straight lines it is the last of the preceding numbers, or six, increased by four, and so on to infinity. These numbers, as is obvious, are triangular and the side of any one of them is the number of points of intersection, which are found on any one straight line, i.e. the number which is one less than the number of all the straight lines. Therefore if the number of all the given points, which is the same as the number of points of intersection on any one straight line, is taken away from this number of all points of intersection, the number left will still be a triangular number, whose side is smaller by one than the side of the previous number, which shows the number of all the points, and hence is two less than the number of straight lines proposed. And this is the number of points of intersection which *Pappus* requires to lie on a straight line which is given in position in this Proposition, and he asserts that if no three of them are at the angles of a triangle, [no four of them are at the angles of a quadrilateral ^{c} and so on], then each one of the remaining points of intersection will lie on a straight line which is given in position. Now we have been forced by necessity to add these things to the text which are enclosed in brackets, for without them the Proposition would not be true beyond the case of five lines.

In fact the Proposition is conveniently divided into two cases; these are also quite clearly indicated by *Pappus*, who states the hypothesis of the simpler case of the Proposition, of this most general and at the same time elegant type.

First Case

Suppose that any number of straight lines meet each other and there are no more than two through the same point; let all the points on one of them be given, and moreover let each individual point on another lie on a straight line which is given in position; each of the remaining points of intersection will lie on a straight line which is given in position. For let there be any number of straight lines, e.g. six, AF, BG, CH, DK, EL, EA; let all the points on one of them, namely A, B, C, D, E, be given; and let all the points on another, namely F, L, M, N lie on a straight line which is given in position: each one of the remaining points of intersection will lie on a straight line which is given in position.

For let any one of the remaining points be taken, e.g. O, and since there are four straight lines OL, ON, AN, AB, and three points are given on one of them, namely A, B, E, and indeed the remaining points apart from O, viz. L, N, lie on a straight line which is given in position, the point O will also lie on a straight line which is given in position by *Proposition* 1. In the same way the same thing will be shown about all of the remaining points.

---

^{c} Here figures whose sides cut across each other like diagonals are to be considered just like the others.

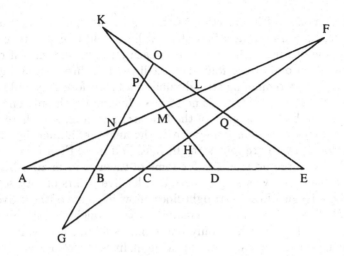

Second Case

With the other things remaining as before, now suppose that all of the points lying on a straight line which is given in position (the number of which is two less than the number of straight lines proposed) are not on the same straight line, but that none of them form an orbit; it has to be shown that all the remaining points lie on a straight line which is given in position.

Lemma 1

Suppose that any number of straight lines meet each other and there are no more than two through the same point, and let any collection of these straight lines be taken; then if two points of intersection are taken on each of the chosen straight lines and the number of points of intersection obtained in this way is equal to the number of these straight lines, these points will form an orbit.

For since there are two points on each individual straight line, there will be at least three on two straight lines, and four on three, and so on; that is to say, the number of points will always be at least one more than the number of straight lines unless the last straight line passes through the first point; i.e. unless the straight lines form an orbit, in which case alone the number of points will be equal to the number of straight lines.

Lemma 2

Suppose that any number of straight lines meet each other and there are no more than two through the same point, and let any of their points of intersection be taken, whose number is equal to the number of all the straight lines; either all these points of intersection, or some of them, will form an orbit, in other words they will be found at the angles of a polygon or a triangle.

For three points of intersection of three straight lines are at the angles of a triangle; then if there are four straight lines, and four points are taken, necessarily one of these will be found on each straight line; now if only one point is found on one of the four straight lines, the three remaining points will be on the three remaining straight lines, and therefore they will be at the angles of a triangle: but if there is no straight line on which only one point is found, there will be two on each of the four straight lines, and there are four points, therefore by *Lemma* 1 they are at the angles of a quadrilateral. And it is clear that if there are four straight lines and more than four points are taken, there are more possibilities for some of them to form an orbit.

Now let there be five straight lines, and let five points of intersection be taken; if there is one of these straight lines on which none of these five points is found, all five will be on the remaining four straight lines; and if there is some straight line on which only one point is found, the remaining four points will be on the remaining four straight lines; therefore in both cases some points will be at the angles of a triangle, or a quadrilateral, by the preceding case: but if there is no straight line on which either no point or a single point is found, there will be two on each of the five straight lines, and there are five points, therefore by *Lemma* 1 they are at the angles of a pentagon. It will be shown for six straight lines by the same straightforward reasoning, and so on to infinity.

In the figure below there are eight straight lines, and eight points are taken, four of which form an orbit.

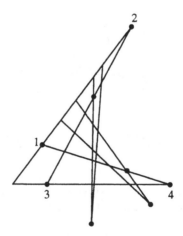

These things having been set forth the Proposition is demonstrated in the following way.

First let there be five straight lines AD, AE, BF, CG, DH, and when the given points on one of the straight lines have been removed, viz. A, B, C, D, six points E, F, G, H, K, L will be left on four straight lines, and let three of them (for the side of the triangular number 6 is 3) which are not at the angles of a triangle contained by three of the proposed straight lines, e.g. E,

F, G, lie on a straight line which is given in position; it has to be shown that the remaining three points K, H, L also lie on a straight line which is given in position.

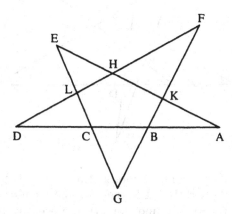

Therefore since there are four straight lines AE, BF, CG, DF, and three points of intersection are taken on them, viz. E, F, G, there will be some one of these straight lines on which necessarily only one of these three points will be found; for otherwise either there will be some straight line on which there is no point, and then there will be three points on the three remaining straight lines, i.e. at the angles of a triangle, contrary to hypothesis, or there will be at least two points on each of the four straight lines, and therefore there would be at least four points; but there are only three; hence it is necessary that there is some straight line on which only one point is found: let this straight line be AE, on which there is the point E, therefore the remaining two F, G are on the remaining three straight lines BF, CG, DF; therefore, since the three points B, C, D are given, the remaining point L on those three straight lines lies on a straight line which is given in position by the first Proposition: now let GE be taken, namely the straight line from these three which passes through the point E on the fourth line, and all points on this straight line GE will lie on a straight line which is given in position. Hence, by the first case of this Proposition, the remaining points K, H lie on a straight line which is given in position.

Now let there be six straight lines AE, AF, BG, CH, DK, EL; and when the five given points A, B, C, D, E, which are on one of the straight lines, have been removed, ten points F, G, H, K, L, M, N, O, P, Q will be left on five straight lines; and by hypothesis four of them, which do not form an orbit, lie on a straight line which is given in position; let these be F, G, H, K; and it has to be shown that the remaining six L, M, N, O, P, Q lie on a straight line which is given in position.

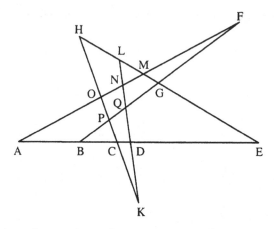

Therefore since four points of intersection F, G, H, K are taken on five straight lines AF, BG, CH, DK, EL, there will be some straight line on which only one of these points is found; for otherwise either there will be some straight line on which there is no point, and then the four points will be on the four remaining straight lines, and therefore some of them will form an orbit by *Lemma* 2, contrary to hypothesis, or there will be at least two points on each of the five straight lines, and so there would be at least five points; but there are only four, so that it is necessary that there is some straight line on which only one point is found; let this be AF on which there is the point F; therefore the remaining three points G, H, K are on the remaining four straight lines BG, CH, DK, EL, and the points B, C, D, E are given; therefore by the first part of this demonstration the remaining three points on these four straight lines, namely L, P, Q, lie on a straight line which is given in position. Now let BF be taken, namely the straight line from these four, which passes through the point F on the fifth straight line; and all points on this straight line BF will lie on a straight line which is given in position: hence by the first case of this Proposition the remaining points M, N, O lie on a straight line which is given in position. In the same straightforward way the Proposition will be demonstrated for seven straight lines, eight straight lines, and so on to infinity, as is clear.

That the condition contained in brackets in this Proposition is entirely necessary, is clear in these two examples; in fact the same thing can be demonstrated generally by means of the preceding material.

The most distinguished Professor has added to these things the following two Porisms, also restored by himself, from the first Book of Euclid's *Porisms.*

The first Porism, Book 1 of *Euclid's* Porisms, which *Pappus of Alexandria* preserved in the preface to *Book* 7 of the Mathematical Collections. *See p. 35 of its Preface.*

If two straight lines are inflected from two given points to a straight line which is given in position, and moreover one of them cuts off from a straight line which is given in position a segment adjacent to a given point on it, the other will also take off from another straight line a segment having a given ratio.

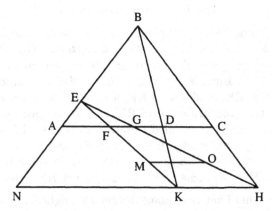

For let two points D, C be given, from which DB, CB are inflected to AB which is given in position; let one of them DB cut off from the straight line EF which is given in position the segment KM adjacent to the given point M: it has to be shown that the other one CB takes off from a certain other straight line a segment having a given ratio to the segment KM.

Let the join CD meet AB, EF which are given in position in the points A, F, which will consequently be given. From the point K, in which the inflected line BD meets the straight line EF, let KH be drawn parallel to AD, and meeting the other inflected line BC in H, and BA in N. Therefore since the points A, D, C are given, the ratio AD to DC and therefore the ratio NK to KH will be given; hence if EH is joined, meeting AD in G, the ratio AF to FG will be given; but AF is given, so that FG is also given, as is the point G; and E is given, so that EG is given in position and magnitude; and EF is given, so that the ratio EF to EG is given; and if MO is drawn through the given point M parallel to AD, and meeting EG in O, then MO will be given in position, and so the point O will be given; and because the straight lines MO, FG, KH are parallel, MK is to OH as EF is to EG, which are in a given ratio. Therefore the straight line BC takes off from the straight line EG which is given in position the segment OH adjacent to the given point O, which is in a given ratio to the segment MK. Q.E.D.

Now it will be composed as follows; let AF be made to FG as AD is to DC, and, EG having been joined, let MO be drawn through M parallel to AD; it has to be shown that, if any straight lines DB, CB are inflected from the points D, C to AB, cutting off from the straight lines EF, EG segments MK, OH adjacent to the points M, O, these will be in the given ratio EF to EG, or, what is the same, the join HK is parallel to AD; now this seems to have been omitted by *Euclid*, possibly because it can be demonstrated briefly in an indirect manner; however *Pappus* provides two direct demonstrations of the same in *Lemma* 1 for the Porisms; we will add below the second of these, which is a little corrupted in *Commandino's* edition, but is here restored to its proper form. (*See* Pappus's *Book 7, folio 239, first page.*)

It may be shown by compound proportion in the following way.

(*See* Pappus's *figure, folio 238, second page, or our immediately preceding figure.*) Since AD is to DC as AF is to FG, *convertendo*, CD will be to DA as GF is to FA, and *componendo*, *permutando* and *convertendo* AC will be to CG as AD is to DF. But the proportion AD to DF is compounded from the proportion AB to BE, [and EK to KF, and the proportion AC to CG is compounded from the proportion AB to BE] and the proportion EH to HG. Therefore the proportion compounded from AB to BE and EK to KF is the same as that which is compounded from AB to BE and EH to HG. Let the ratio AB to BE be removed from both proportions, therefore the remaining proportion EK to KF is the same as EH to HG; hence HK is parallel to AG.

The Second Porism. That that point lies on a straight line which is given in position.

The second Porism seems to require to be explained in the following way.

Suppose that from two given points C, G two straight lines CB, GD are drawn, meeting two straight lines AB, ED which are given in position, and that the straight line DB joining the points of intersection is parallel to the straight line CG, which is drawn through the given points; the point of intersection K of the drawn lines will lie on a straight line which is given in position.

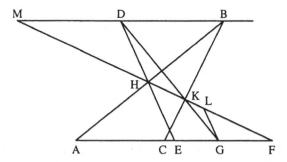

For let the straight lines which are given in position meet each other in H, and let the join KH meet CG in F and BD in M. Therefore because of the parallels AE is to EF (as BD is to DM that is) as CG is to GF; and therefore AE is to CG as EF is to GF, and so the ratio EF to GF is given, and EG is given, so that the point F is given, and the point H is given, so that HF is given in position.

And so let CG be to GF as AE is to EF, and the join HF will be the straight line on which the point K lies; that is to say, if any straight line GD is drawn, meeting FM in K, there will be a straight line which passes through C, K, B, for DB is to DM as AE is to EF, that is by construction as CG is to GF, so that DB is to CG as (DM is to GF, that is as) DK is to KG, and therefore CKB is a straight line.

Pappus demonstrates the same thing in another way in the second *Lemma*, which must be read in the following way, namely: (*See* Pappus's *figure, folio 239, second page, or our last figure above.*) Through G let the straight line GL be drawn parallel to DE, and let the join HK be produced to L. Now CG is to GF as AE is to EF, and *permutando* [AE is to CG as EF is to GF]; but EH is to GL as AE is to CG, because two straight lines are parallel to two straight lines. [d] Therefore EH is to GL as EF is to FG, and EH is parallel to GL, therefore there is a straight line which passes through H, K, L, F.

Some Notes on Simson's Demonstrations

The demonstrations of the two general Propositions presented in the *Treatise* are a little more detailed than those in the paper and so further clarification may be obtained by consulting Propositions 19 and 21 and their notes. However, it may be useful to draw attention to the following points concerning the demonstrations of these Propositions and the first Porism.

(i) At the end of the demonstration of the first Proposition Simson asserts from $\frac{HL}{LG} = \frac{MF}{FN}$ that $\frac{HG}{MN} = \frac{HK}{MK} = \frac{HL}{MF}$. Referring to the first diagram we deduce first that $\frac{HL}{LG-HL} = \frac{MF}{FN-MF}$, i.e. $\frac{HL}{HG} = \frac{MF}{MN}$, so that $\frac{HL}{MF} = \frac{HG}{MN} = \frac{HK}{MK}$, the last equality coming from the similar triangles GHK and MKN. In the case of the second diagram we argue similarly but start with $\frac{HL}{LG+HL} = \frac{MF}{FN+MF}$.

(ii) In the demonstration of the second case of the second Proposition for five straight lines the first Proposition is applied to the four straight lines BD, DF, FG, GL; the points B, C, D are given points on BD, while each of the points F, G on FG lies on a straight line which is given in position, so that the point L also lies on a straight line which is given in position.

In the demonstration for six straight lines we consider the five straight lines BE, EH, HK, KL, BG for which the points B, C, D, E on BE are all given, while each of the points G, H, K lies on a straight line which is given in position and they are not at the vertices of a triangle formed by any three of the five straight lines; therefore by the result for five straight lines each of the remaining points of intersection for these five straight lines, L, P, Q, lies on a straight line which is given in position.

(iii) For the first Porism we use $\frac{MK}{OH} = \frac{EF}{EG}$. This comes from $\frac{EF}{EG} = \frac{EM}{EO} = \frac{EK}{EH} = \frac{EK-EM}{EH-EO} = \frac{MK}{OH}$.

(iv) In the *compound proportion* demonstration application of Menelaus's theorem to triangle AEF and transversal BDK gives $\frac{AB}{BE} \cdot \frac{EK}{KF} \cdot \frac{FD}{DA} = 1$, or

---

[d] *Because two straight lines are parallel to two straight lines,* namely AE to DB, and GL to DE, from which AE is to DB as EH is to DH, and DB is to CG (as DK is to KG, i.e.) as DH is to LG; therefore by [V 22] AE is to CG as EH is to GL.

$\frac{AD}{DF} = \frac{AB}{BE} \cdot \frac{EK}{KF}$; from triangle AEG and transversal BCH we obtain similarly
$\frac{AC}{CG} = \frac{AB}{BE} \cdot \frac{EH}{HG}$.

Simson's Two Interpretations of the First Porism

In the above discussion of the first Porism Simson requires that the ratio of OH to MK is independent of the position of B on the straight line AB, which is given in position. However the actual ratio is not specified and Simson notes in the *Treatise* (paragraph following the enunciation of Proposition 23) that he subsequently found "innumerable mutually parallel straight lines which satisfy the Porism proposed in this way". Presumably he had in mind the following configuration where, with reference to the above figure for the first Porism, the straight line l is parallel to EG and meets BC in Y and OC in X.

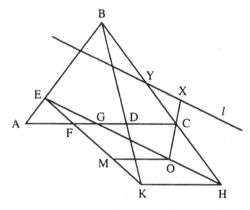

Now X is the fixed point on l and XY is the segment cut off by CB. We have $\frac{XY}{OH} = \frac{XC}{CO}$, which is fixed for l, and so $\frac{XY}{MK} = \frac{XY}{OH} \cdot \frac{OH}{MK}$ is fixed for l. Thus the required condition is satisfied for l, although clearly the ratio $\frac{XY}{MK}$ depends on the particular line l chosen parallel to EG. In the *Treatise* Simson looks at the situation where the ratio is also specified. The problem now is to find the particular lines l which give $\frac{XY}{MK}$ the specified value. As Simson notes, there are in fact two such lines.

A2. "That this goes to a given point"

In Proposition 34 Simson gave a reconstruction of this, the sixth Porism from Euclid's first Book according to Pappus. Simson also recorded other results under this heading. For example, in his letter to Dr Jurin of 10 January 1724 he wrote:[28]

"I shall subjoin the following Porism which I have the best Grounds to think one of Euclid's own.

That this goes to a given point

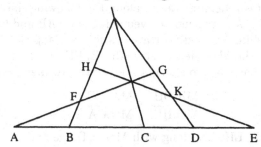

Suppose that five points A, B, C, D, E have been given on a straight line. Through any one of them C let a straight line be drawn in any manner and from any two of the remaining points A, E let two straight lines be inclined to the drawn line and likewise from the other two points B, D let two straight lines be inclined to the same straight line and suppose that they meet the first two in four points F, G, H, K. If any opposite pair of these points, F, K or H, G, is joined by a straight line, then this will go to a given point.

I would have added the demonstration, but am so much wearied with the length of this Letter I must leave it to another occasion."

The following generalised version along with an indication of Simson's demonstration appears in Volume D of the *Adversaria* [43, p. 110 (see also p. 100)] – again the statements are in Latin.

Porism [That this will go to a given point] for which Propositions 129, 130 of Pappus's Book 7 are useful.

General Case

From four given points A, B, C, D on the same straight line let four straight lines be drawn which meet each other in six points E, F, G, H, K, L and suppose that the straight line EH joining one of the pairs of points which are not on the same one of the drawn lines goes to a given point on AB. Then the straight line through any two of the remaining points, for example the straight line FL through F and L, will also go to a given point.

---

[28]The Porism was enunciated in Latin in the letter. See [47].

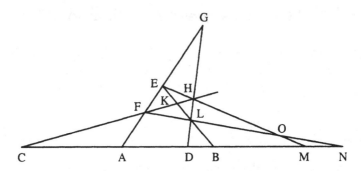

Simson indicates a demonstration along the following lines. Suppose, for example, that EH passes through a given point M on AB and let FL meet AB in N and EH in O. Simson makes two applications of Pappus's Proposition 129 (Proposition 17 in the *Treatise*). First, with FE, FH, FN as the three straight lines and ME, MC as the two straight lines drawn to them, we obtain

$$\frac{ME.OH}{MO.HE} = \frac{MA.NC}{MN.CA}.$$

Secondly, using LE, LH, LN along with ME, MD, we find

$$\frac{ME.OH}{MO.HE} = \frac{MB.ND}{MN.DB}.$$

Thus

$$\frac{MA.NC}{MN.CA} = \frac{MB.ND}{MN.DB}, \quad \text{so that} \quad \frac{NC}{ND} = \frac{MB.CA}{MA.DB}.$$

Since the ratio $\frac{NC}{ND}$ is therefore given and the points C and D are given, it follows that the point N is given.

We see similarly that GK passes through a given point on AB.

A3. Correspondence between Pappus's Lemmas and Simson's Propositions

All of Pappus's Lemmas for the Porisms are discussed by Simson. Individual Lemmas are often given in extended form and sometimes appear within Simson's discussion of a Proposition. The following table shows where each of the Lemmas can be found in Simson's text.

Pappus Lemma	Prop.	Simson Prop.	Pappus Lemma	Prop.	Simson Prop.
1	127	22	20	146	69
2	128	8 Case 4 Art. 2	21	147	70
3	129	17	22	148	42
4	130	18	23	149	43
5	131	13	24	150	44
6	132	8 Case 2	25	151	45
7	133	8 Case 3	26	152	71
8	134	9	27	153	72
9	135	8 Case 5	28	154	73
10	136	26	29	155	49
11	137	28	30	156	51
12	138	29	31	157	74
13	139	30	32	158	75
14	140	31	33	159	63
15	141	32	34	160	64
16	142	27	35	161	65
17	143	33	36	162	76
18	144	39	37	163	77
19	145	68	38	164	79

A4. Corrections to Simson's Text

In the course of this work I have identified and corrected a number of errors in Simson's text. These are listed below for the benefit of anyone who may wish to consult the original work. The initial page and line numbers refer to Simson's text. The corresponding locations in *Simson on Porisms* appear within brackets following the correction.

1. (p. 319, l. 12) Insert *intellecta* between *mihi* and *nec*. (p. 14, l. 17)
2. (p. 343, l. 18) $\overline{c-d}+c-d \times y$ should be $\overline{c-d} \times x + \overline{c-d} \times y$. (p. 30, l. 26)
3. (p. 344, l. 18) Insert *ipsum* after *Locum*. (p. 31, l. 22)
4. (p. 351, l. 24) *contentum* should be *contento*. (p. 37, l. 15)
5. (p. 355, l. 27) AG should be FG. (p. 49, l. 18)
6. (p. 360, l. 24) *quraum* should be *quarum*. (p. 53, l. 8)
7. (p. 372, diagram) H should be M. (p. 61)
8. (p. 372, l. 2) FR should be FK. (p. 62, l. 12)
9. (p. 374, l. 16) *reliquum* should be *reliquam*. (p. 63, l. 24)
10. (p. 385, l. 8) *datum* should be *data*. (p. 71, l. 25)
11. (p. 408, l. 1) BEZ should be BNZ (since at this point E has not been defined). (p. 89, l. 41)
12. (p. 417, l. 7) CD should be ED. (p. 105, l. 15)
13. (p. 418, l. 7) D should be E. (p. 106, l. 2)
14. (p. 427, l. 1) 26 should be 28. (p. 112, l. 29)
15. (p. 428, l. 6) The bracket) should come after *est* and not *rationi*. (p. 113, l. 25)
16. (p. 429, ll. 15, 23) 3 should be 2. (p. 114, ll. 14, 24)
17. (p. 472, l. 20) 59 should be 61. (p. 145, l. 32)
18. (pp. 482, 484, diagram) D, G should be interchanged. (pp. 153 (first diagram), 155 (second diagram))
19. (p. 483, diagram) Y should be V. (pp. 153 (second diagram), 156)
20. (p. 495, ll. 19–21) *inflexa* should be *inflexis* and *occurrentes* should be *occurrentibus* (cf. p. 497, ll. 26–28). (p. 162, ll. 12–15)
21. (p. 495, l. 28) CD, DE should be CB, BE. (p. 162, l. 21)
22. (p. 500, ll. 7-8) *quibus* should be *quo* (two places). (p. 165, ll. 27–28)
23. (p. 529, l. 5) 163 should be 164. (p. 187, l. 4)
24. (p. 530, l. 12) *data* should be *datum*. (p. 201, l. 9)
25. (p. 537, l. 16) CA and DA should be interchanged. (p. 205, l. 24)
26. (p. 545, l. 3) CH should be eH. (p. 210, l. 13)
27. (p. 545, l. 10) bc should be bC. (p. 210, l. 18)
28. (p. 547, l. 29) VD should be Vd. (p. 211, l. 15)
29. (p. 548, l. 27) Kn should be kn. (p. 212, l. 20)
30. (p. 549, l. 2) Cb should be CB. (p. 213, l. 1)
31. (p. 563, l. 14) *occurrunt* should be *occurrit*. (p. 222, l. 10)
32. (p. 574, l. 1) *perpendioularis* should be *perpendicularis*. (p. 229, l. 1)

Also in the Greek quotations on pages 455, 463 and 471 some of the accents are wrong. I believe that they are set correctly on the corresponding pages 133, 139 and 145 of *Simson on Porisms*.

References

1. Monographs and Journal articles

1. Ismael Boulliau: Ismaelis Bullialdi exercitationes geometricae tres. I. Circa demonstrationes per inscriptas & circumscriptas figuras. II. Circa conicarum sectionum quasdam propositiones. III. De Porismatibus. Astronomiae Philolaicae fundamenta clarius explicata, & asserta aduersus clariss. viri Sethi Wardi ...impugnationem. Parisiis, 1657.
2. Henry Brougham (Baron Brougham and Vaux): Lives of Philosophers of the Time of George III. 1855.
3. Michel Chasles: Les trois livres de Porismes d'Euclide, rétablis pour la première fois, d'après la notice et les lemmes de Pappus, et conformément au sentiment de R. Simson sur la forme des énoncés de ces propositions. Paris, 1860.
4. Federicus Commandinus Urbinas: Apollonii Conicorum Libri Quarti. Bologna, 1566.
5. Federicus Commandinus Urbinas: Pappi Alexandrini Mathematicae Collectiones. Pesaro, 1588.
6. Leonhard Euler: Variae demonstrationes geometriae. Novi Commentarii Academiae Scientiarum Imperialis Petropolitanae 1, 49–66 ((1747–48), 1750).
7. Leonhard Euler: Opera Omnia, Series I: Opera Mathematica. Lausanne, 1953.
8. Pierre de Fermat: Varia Opera Mathematica .... Accesserunt selectae quaedam ejusdem epistolae, vel ad ipsum a plerisque doctissimis viris ...de rebus ad mathematicas disciplinas aut physicam pertinentes scriptae. Tolosae, 1679.
9. C.C. Gillispie (Editor): Dictionary of Scientific Biography. Scribner's, New York, 1970–1980.
10. Albrecht Girard: Tables des sinus, tangentes & secantes .... Avec un traicté succinct de la trigonométrie, et. (Second Edition) La Haye, 1629.
11. Albrecht Girard: Les oeuvres mathématiques de Simon Stevin. Leiden, 1634.
12. David Gregory: Euclidis quae supersunt omnia. Ex recensione Davidis Gregorii. Oxford, 1703.
13. Edmund Halley: Apollonii Pergaei de Sectione Rationis Libri Duo. Oxford, 1706.
14. Edmund Halley: Apollonii Pergaei Conicorum Libri Octo et Sereni Antissensis De Sectione Cylindri et Coni Libri Duo. Oxford, 1710.
15. T.L. Heath: The Thirteen Books of Euclid's Elements. Cambridge, 1908.
16. T.L. Heath: A Manual of Greek Mathematics. Oxford, 1931.
17. Alexander Jones: Pappus of Alexandria: Book 7 of the Collection (Sources in the History of Mathematics and Physical Sciences, 8 (Parts I & II)). Springer, New York, 1986.
18. John Lawson: A Dissertation on the Geometrical Analysis of the Antients, with a Collection of Theorems and Problems without Solutions. Canterbury, 1774.
19. Thomas Leybourn (Editor): The Mathematical Repository (Second Edition). London, 1799–1804.

20. J.S. Mackay: Mathematical Correspondence – Robert Simson, Matthew Stewart, James Stirling. Proc. Edinburgh Math. Soc. **21**, 2–39 (1903).
21. Colin MacLaurin: A letter from Mr. Colin Maclaurin, Math. Prof. Edinburgh, F. R. S. to Mr. John Machin, Astr. Prof. Gresh. & Sec. R. S. concerning the Description of Curve Lines. Phil. Trans. **39**, 143–165 (1735–36).
22. John Playfair: On the origin and investigation of Porisms. Trans. Roy. Soc. Edinburgh **3**, 154–204 (1794).
23. Robert Potts: Simson's Restoration of the Porisms. Mathematical Tracts 1843–1854, University of London Library (L°(B.P. 38) De Morgan Library).
24. Carlo Renaldini (Count): Carlo Renaldini ... de resolutione et compositione mathematica ... libri II. Patavii, 1668.
25. August Richter: Porismen nach R. Simson bearbeitet und vermehrt, nebst den Lemmen des Pappus zu den Porismen des Euklides. Elbing, 1837.
26. Robert Simson: Pappi Alexandrini Propositiones duae generales, quibus plura ex Euclidis Porismatis complexus est, Restitutae a Viro Doctissimo Rob. Simson, Math. Prof. Glasc. Vid. Pappi praefationem ad Lib. 7. Coll. Math. Apollonii de Sectione rationis libris duobus a Clariss. Hallejo praemissam pag. VIII & XXXIV. Phil. Trans. **32**, 330–340 (1723).
27. Robert Simson: Apollonii Pergaei Locorum Planorum Libri II. Glasgow, 1749.
28. Robert Simson: Sectionum conicarum libri V (Second Edition). Edinburgh, 1750.
29. Robert Simson: Euclidis Elementorum Libri priores sex, idem undecimus et duodecimus. Glasgow, 1756.
30. Robert Simson: The Elements of Euclid. Glasgow, 1756.
31. Robert Simson: The Elements of Euclid (Second Edition), with The Book of Euclid's Data. Glasgow, 1762.
32. Robert Simson: Opera Quaedam Reliqua. Glasgow, 1776. [Editor: James Clow.]
33. Robert Simson: Apollonii Pergaei de Sectione Determinata Libri II restituti, duobus insuper libris aucti. Opera Quaedam Reliqua, 1–313.
34. Robert Simson: De Porismatibus Tractatus. Opera Quaedam Reliqua, 316–594.
35. Robert Simson: A Treatise concerning Porisms, translated from the Latin by John Lawson B. D. Canterbury, 1777.
36. Matthew Stewart: General Theorems of considerable use in the higher parts of Mathematics. Edinburgh, 1746.
37. Matthew Stewart: Pappi Alexandrini collectionum mathematicarum libri quarti proposito quarta generalior facta, cui propositiones aliquot eodem spectantes adjiciuntur. Edinburgh Physical Essays **1**, 141–172 (1754).
38. Matthew Stewart: Propositiones Geometricae more veterum demonstratae. Edinburgh, 1763.
39. William Trail: Elements of Algebra for the use of Students in Universities. Aberdeen, 1770.
40. William Trail: Account of the Life and Writings of Robert Simson, M.D. London, 1812.
41. A.P. Treweek: Pappus of Alexandria: the manuscript tradition of the *Collectio Mathematica*. Scriptorum **11**, 195–233 (1957).
42. D.T. Whiteside: The Mathematical Papers of Isaac Newton. Cambridge, 1967–1981.

2. Manuscripts in Glasgow University Library

43. Robert Simson: Adversaria (1716–1767). MS Gen 256–271.
44. Robert Simson: Notae ad Pappi Collectiones. MS Gen 1232.
45. Robert Simson: Annotations in a copy of Pappus Alexandrinus: Mathematical Collections. MS Gen 1118.

46. Correspondence between Robert Simson and Matthew Stewart: MS Gen 146 (see also [20]).

47. Letters from Robert Simson to Dr Jurin. MS Gen 1096. [The Glasgow versions are copies obtained in the 1750s of the originals, which are in the Library of the Royal Society, London (LBC.16.412, 426, 456, 501, 504 and LBC.18.4).]

48. Simson's will. A copy of the will is contained in 'Letters to Professor Robert Simson and other documents' (MS Murray 660). The actual will is available in the Commissary of Glasgow Testaments Vol. 64 (CC9/7/67), Scottish Records Office, Edinburgh.

49. Letters from Francis Hutcheson to Rev. Thomas Drennan. MS Gen 1018/5, 12.

Index

Adversaria   3, 16, 38, 40, 132, 191, 193, 246, 263

Apollonius of Perga   1, 3–4, 19, 21–22, 42, 75, 88, 139, 185, 191, 197–198, 252

Boulliau, Ismael   13, 38
Brougham, Henry (Baron Brougham and Vaux)   2, 4, 40

Chasles, Michel   4
Clow, James   2–3
Commandino, Federico   1, 52, 55, 64, 68, 70, 83–84, 92, 102, 113, 119, 131, 138, 180, 185, 188–189, 197, 260

Drennan, Thomas   3, 4

Euclid   1–5, 13–15, 17–18, 31, 33–35, 38, 42, 45, 59–60, 84, 91–92, 97, 106, 114–115, 133, 139, 145, 165, 175, 187, 189, 191, 251–252, 258, 260, 263
Euler, Leonhard   207, 244

Ferguson, Adam   41
Fermat, Pierre de   13–15, 31–32, 38, 42, 199, 201, 203–205, 207, 210, 217–218, 221, 242–245

Girard, Albert   13, 38
Gregory, David   14, 38

Halley, Edmund   1–2, 11, 14, 39, 43, 139, 185, 197–198, 252
Heath, T.L.   91
Hutcheson, Francis   3
Hyptios Porism   43, 93, 188

Jones, Alexander   4, 6, 36, 41–43, 91, 93, 95, 189
Jurin, James   2, 43, 263

Lawson, John   4
Leybourn, Thomas   4

Machin, John   2
MacLaurin, Colin   3–4, 39, 188, 248
Menelaus of Alexandria   6, 95, 261
Moor, James   39

Newton, Sir Isaac   93, 95

Pappus of Alexandria   1–2, 4, 6, 11, 13–15, 17, 20–21, 24, 31, 33–34, 37, 40–43, 45, 51–52, 55, 58–60, 64, 67–72, 77–78, 82–84, 91–95, 97, 99–106, 111–113, 115, 117–119, 124, 130–131, 133, 137–139, 145, 169–171, 175–180, 183–185, 187–189, 191, 193, 196, 198, 248, 251–252, 254, 258, 260–261, 263–265
Pappus's Proposition on the Four Lines   111, 189
Pappus's Theorem   188
Philosophical Transactions   2–3, 14, 72, 84, 93–94
Playfair, John   41–42
Potts, Robert   4

Renaldini, Carlo (Count)   14, 38
Richter, August   4

Serenus of Antissa (Antinoupolis)   176, 197
Simson, Robert   1–7, 9, 11, 38–43, 45, 91–95, 97, 188–199, 242–249, 251–252, 261–267
Stanhope, Philip (2nd Earl Stanhope)   2
Stevin, Simon   13, 38
Stewart, Dugald   41
Stewart, Matthew   4, 15, 38–41, 145, 199, 217, 228, 232, 238, 248

Trail, William   2–3, 41–42, 135, 192

Whiteside, D.T.   93, 95
Williamson, John   39–40

Sources in the History of
Mathematics and Physical Sciences

Continued from page ii

Hogendijk J.P.
Ibn Al-Haytham's *Completion of the Conics*

Jones A. (Ed.)
Pappus Of Alexandria - Book VII of the Collection

Kheirandish E.
The Arabic Version of Euclid's Optics

Lützen J.
Joseph Liouville 1809-1882

Lützen J.
The Pre-History of the Theory of Distributions

Meyenn K. von, Hermann A., Weisskopf V.F. (Eds)
Wolfgang Pauli: Scientific Correspondence II: 1930-1939

Meyenn K. von (Ed.)
Wolfgang Pauli: Scientific Correspondence III: 1940-1949

Meyenn K. von (Ed.)
Wolfgang Pauli: Scientific Correspondence IV, Part I: 1950-1952

Meyenn K. von (Ed.)
Wolfgang Pauli: Scientific Correspondence IV, Part II: 1953-1954

Moore G.H.
Zermelo's Axiom of Choice: Its Origins, Development, and Influence

Neugebauer O. (Ed.)
Astronomical Cuneiform Texts

Neugebauer O.
A History of Ancient Mathematical Astronomy

Ragep F.J.
Naṣīr Al-Dīn Al-Ṭūsī's *Memoir on Astronomy*

Rosenfeld B.A.
A History of Non-Euclidean Geometry

Sesiano J.
Books IV to VII of Diophantus' Arithmetica: In the Arabic Translation Attributed to Qustā ibn Lūqā

Stephenson B.
Kepler's Physical Astronomy

Swerdlow N.M., Neugebauer O.
Mathematical Astronomy in Copernicus' *De Revolutionibus*

Toomer G.J. (Ed.)
Apollonius: Conics Books V to VII: The Arabic Translation of the Lost Greek Original in the Version of the Banū Mūsā

Toomer G.J. (Ed.)
Diocles on Burning Mirrors: The Arabic Translation of the Lost Greek Original

Truesdell C.
The Tragicomical History of Thermodynamics, 1822-1854